Resource Allocation in Hierarchical Cellular Systems

For a complete listing of the *Artech House Mobile Communications Library*, turn to the back of this book.

Resource Allocation in Hierarchical Cellular Systems

Lauro Ortigoza-Guerrero
A. Hamid Aghvami

Artech House
Boston • London

Library of Congress Cataloging-in-Publication Data
Ortigoza-Guerrero, Lauro.
   Resource allocation in hierarchical cellular systems / Lauro Ortigoza-Guerrero, A. Hamid Aghvami.
        p.   cm. — (Artech House mobile communications library)
   Includes bibliographical references and index.
   ISBN 1-58053-066-4 (alk. paper)
   1. Radio frequency allocation.   2. Cellular radio.   I. Aghvami, A. Hamid.   II. Title
III. Series.
   HE8675.O78   1999
   384.54'524—dc21   99-045831
                                                                            CIP

British Library Cataloguing in Publication Data
Ortigoza-Guerrero, Lauro
   Resource allocation in hierarchical cellular systems. — (Artech House mobile
   communications library)
   1. Wireless communication systems   2. Radio frequency allocation
   I. Title   II. Aghvami, A. Hamid
   621.3'8456

   ISBN 1-58053-066-4

Cover design by Andrew P. Ross

© 2000 ARTECH HOUSE, INC.
685 Canton Street
Norwood, MA 02062

International Standard Book Number: 1-58053-066-4
Library of Congress Catalog Card Number: 99-045831

10 9 8 7 6 5 4 3 2 1

Dedicated to our families

Contents

Preface

This book is intended to help cellular mobile telephone network operators, telecommunications engineers, cellular planners, telecommunications students and teachers, and researchers in the area of mobile cellular systems in general understand the basic ideas behind channel allocation strategies and their application to cellular systems.

The idea is to present the information as clearly and simply as possible and gradually build in complexity, starting with an explanation of channel allocation strategies and their classification and then their application to single layered systems and its mathematical treatment. This leads to a similar treatment of hierarchical cellular systems (HCSs), which are likely to dominate future worldwide multimedia systems. Wherever possible, throughout the book the systems considered are those proposed for use in universal mobile telecommunication systems (UMTS), so the performances of real systems are analyzed and simulated. In addition, care has been taken to use traffic patterns obtained from actual working systems to add reality to the performances obtained.

The core of the material is divided in three broad areas:

- Application of resource allocation strategies to single-layered cellular systems;
- Spectrum partitioning for HCSs;
- Resource allocation strategies for HCSs.

The book starts by introducing the idea of channel allocation strategy and presenting their classification based on how they are implemented. Simple examples are presented of channel assignment strategies being applied to micro-cellular systems—first formed by hexagonal shaped cells and the later by Manhattan-like cells. Their performances are assessed by means of a system-level simulator.

The concept of spectrum efficiency and cell capacity is then introduced and calculated for the two main multiple access schemes (MASs) proposed for UMTS: wide-band time division multiple access scheme/code division multiple access scheme (WB-TD/CDMA) and wide-band code division multiple access scheme (WB-CDMA). Then the book deals with spectrum partitioning for a two-layer HCS and analyzes and simulates some resource allocation strategies with HCSs using wide-band time division multiple access scheme/code division multiple access scheme with frequency division duplex (WB-TD/CDMA-FDD), as defined for UMTS. As in the rest of the book, MASs proposed for UMTS have been used, whenever possible, to help find out their spectrum efficiency.

We sincerely thank all the many people who have assisted us in the preparation of this book, especially Mr. J. Pearson, whose suggestions and comments were invaluable in preparing this text. We still and always will remember the valuable advice, helpful discussions, and encouragement we received from all our colleagues at the Centre for Telecommunications Research during the course of the preparation of the new material included here.

We also offer our deepest appreciation and gratitude to our families for their unfailing support and encouragement during the preparation of this book because, as any author knows, they were inevitable greatly affected by the time it all takes. Finally, financial support from the Consejo Nacional de Ciencia y Tecnología (CONACyT), México for much of the work involved, is gratefully acknowledged.

Lauro Ortigoza-Guerrero and Hamid Aghvami
Centre for Telecommunications Research
London 1999

1

Introduction

The objective of this book is to present an up-to-date survey and description of efficient channel allocation strategies (CASs) for use in future wireless cellular mobile telecommunication systems (WCMTSs) formed by hierarchical cellular structures (HCSs).

1.1 Overall Descriptive View of Contents

This book aims to present up-to-date information on a topic of great interest to research scientist and engineers working on mobile communication systems. The topic is that of WCMTS design, both now and in the future, when they will require efficient CASs and HCS. It is based on work carried out for a Ph.D. degree at King's College London.

While care is taken to provide adequate descriptive sections throughout the book, its contents are deliberately intended to present the mathematical theory behind the designs and to give relevant system performance results obtained both analytically and by computer simulation from the results of the theory. Therefore, it is hoped that the book will cater to both readers with a general telecommunications background who want to know what is going on and the more specialist design engineers who need to know and understand the tools and methods being applied.

1.1.1 The Need for WCMTS

The radio spectrum allocated to these networks is fixed and scarce, and efforts have to be made to optimize its utilization. Efficient use of the available radio spectrum in a WCMTS will produce an increase in the number of users that can be attended by the system and/or an increase in the number of calls that terminate satisfactorily without interruption during their lifetime. It is also important to manage the available radio resources efficiently from the cost-of-service point of view. All these factors decrease the costs of the infrastructure required to deploy or extend such a network since it causes a reduction in the number of base stations (BSs) required to give service to a particular geographical area.

1.1.2 Basic Principles of WCMTS Operation

The fundamental operational principle of a WCMTS is the reuse of frequencies at different places within the area of service. The most prohibiting factor in reusing the frequencies is the interference caused by the environment or by other mobiles. One way of decreasing this interference is to use CASs, whose main idea is to use radio propagation path loss to minimize the cochannel interference caused by frequency reuse, so that, when far away from the transmitter, reusing the same frequency will not interfere, as it will be attenuated enough. In doing so, the carrier-to-interference ratio (CIR) and the radio spectrum reuse efficiency are improved.

In general terms, a HCS may consist of two or more layers of cells, with the layer having the smallest cells the lowest in the hierarchy. They may be formed of a picocellular layer to give service to indoor environments and a microcellular layer to provide service to indoor/outdoor environments. Both could be overlaid by a macrocellular layer that deals with requests from users in outdoor areas in cities or rural areas. Finally, at the highest hierarchical level, there could be communication satellite beams overlaying all the previous terrestrial layers or some clusters of them. Small cells provide greater spectral reuse and larger capacity and allow the use of low-power lightweight handheld mobile user devices. They also provide coverage to points that larger cells cannot reach due to natural or man-made obstacles. Large cells are used to cover larger areas with low-cost implementation, to cover spots that are difficult for radio propagation in small cells, and to provide overflow groups of channels for clusters of small cells when heavily loaded. Figure 1.1 shows an example of a four-level HCS.

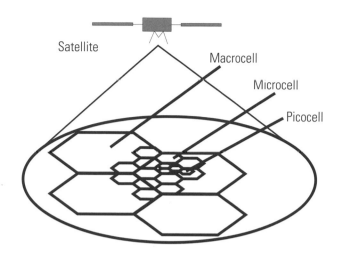

Figure 1.1 Four-level HCS.

1.1.3 HCSs

HCSs are used in WCMTSs to help improve coverage, to increase capacity, to balance the load between layers, and to give service to users with different mobility characteristics (i.e., slow or fast). The speed of the user determines which cellular layer is the most adequate to provide service. Once a layer is selected, efforts are made to attend the user in that layer. Where there is a lack of resources in the appropriate layer, the next higher layer in the hierarchy with larger cells would attend the call.

It is common to associate people working in indoor environments with picocells, pedestrians and users in cars traveling at low speeds with microcells, users in cars moving at high speeds with macrocells, and users traveling in ships and airplanes with the highest levels of the hierarchy or satellite. However, there is no clear classification of when a user should be considered as a slow-moving user or a fast-moving user. The threshold to differentiate one from the other can even depend on the offered load, and can change during the day. Thresholds are normally proposed by system operators to differentiate users and depend mainly on the probability density function of their speed, the time required to establish a call or a handover, and the cells' radii. Successful classification of users in a HCS in terms of their speed helps to reduce the forced termination of calls in progress in layers with small cells and the signaling load between base station controllers (BSCs) and to improve the use of channels in layers with large cells when acting as overflow servers. The classification of users most commonly adopted in a HCS formed of two layers, microcells and macrocells, is the one that considers only slow- and fast-moving users with the division between the two decided by operational realities, as below.

The type of traffic users generate in any particular layer is different in nature than that generated by users in another. In the early stages of WCMTSs, the ratio of data users to speech users was low and the offered traffic was mainly generated by speech users, but with the migration of many applications to multimedia platforms, the ratio has risen to a point that it is almost unity and keeps growing. In the past, data users were normally associated with layers having small cells but now behave in a way similar to speech users, being present at all hierarchical levels.

1.1.4 Radio Resource Management in HCS

A CAS does not make efficient use of the radio resources only by maximizing the CIR but by managing them within and between the cells of the layers, considering both time and space distributions of traffic. This task is also fulfilled by coping with different user bandwidth needs and by making the best of multiple access schemes (MASs), and their combinations, and of duplexing techniques to deal with asymmetric traffic loads such as the internet.

When the same band of frequencies has to be shared between different layers of a HCS, spectrum partitioning becomes a big issue. The bandwidth division is made to take into account the speed characteristics of users and the traffic expected in each layer. Obviously, the bandwidth partitioning is fixed, and therefore there can be no flexible sharing of resources unless a CAS is used.

1.2 Overall View of Technical Content of the Book

A brief description of the technical content is provided in this section. A summary for every chapter of the book is presented to give readers an idea of what they may find. The relationship between the content of every chapter with each other is described. Also, the organization of the book is explained.

1.2.1 Channel Allocation Strategies

In Chapter 2, a general summary is given of current CASs and how they are likely to develop in the future. Ideas are presented of what a CAS is and how it is used. Then CAS classification methods are discussed and a classification is introduced based on the way in which cochannels are distributed. Hence, this divides the strategies into fixed channel allocation (FCA), dynamic channel allocation (DCA), and hybrid channel allocation (HCA) strategies. The advantages and disadvantages of these strategies are outlined, and examples of each are given. Other types of classification are also mentioned and discussed briefly.

1.2.2 Performance Analysis Methods for a Single-Layer Microcellular System

In Chapter 3, using a simplified traffic model based on real data, methods are explained for the mathematical analysis of the performance of a basic cellular system consisting of a single layer with several cell clusters.

The model uses a FCA strategy for initial access with prioritized and nonprioritized handover procedures. The prioritized handover procedure considered is the reserved channel scheme (RCS), in which a set of channels is reserved for handover calls only, whereas in the no priority scheme (NPS) no channels are reserved. Uniform and nonuniform traffic distributions are used and their performances verified by computer simulations. In addition, a method is derived of assessing handover arrival rates for any kind of traffic based on knowledge of new call arrival rates in each cell. Finally a modified version of the DCA strategy called compact pattern with maximized channel borrowing (CPMCB) is described. CPMCB is a strategy that gives prioritization to handover calls. This leads to either a reduction in infrastructure or an increase in capacity for the same grade of service.

1.2.3 MASs for Universal Mobile Telecommunication Schemes in Manhattan-like Environments

In a rather more complex microcellular environment consisting of a rectangular grid of intersecting streets, normally referred to as Manhattan-like microcells, the capacity of the two main MASs proposed for universal mobile telecommunication systems (UMTS) is assessed in Chapter 4. The capacity assessment is done by system level simulations based on results produced by link layer simulations. The MASs taken into consideration are the hybrid wideband time division multiple access scheme/code division multiple access scheme with timedivision duplex (WB-TD/CDMA-TDD), and the wideband code division multiple access scheme with frequency division duplex (WB-CDMA-FDD). They are evaluated in combination with CASs. The former system uses a very simple DCA strategy and the later one the classic FCA strategy.

1.2.4   HCS for WCMTS

Before discussing the design of a DCA for HCS, spectrum partitioning for a two-level HCS WCMTS using an analytical method is addressed in Chapter 5. In a two-level HCS, the available spectrum must be divided between them in a dynamic manner to make best use of the resources. The spectrum partitioning is calculated based on the probability that a user is blocked from each layer (macrocellular and microcellular) when trying to establish a communication. The technique alters spectrum partitioning between the two layers and allows overflow of traffic from microcells to macrocells to maintain an adequate grade of service. FCA is used in both layers and the speed of the user is taken into account; verification is via computer simulation.

1.2.5   A Distributed Dynamic Resource Allocation Strategy for HCS

In Chapter 6, a distributed dynamic resource allocation (DDRA) strategy for a simple HCS is presented. The HCS is formed of a microcell layer overlaid by a macrocell layer. The users are identified as fast- and slow-moving users according to their speeds. Based on this classification, they are directed to an optimum layer: fast-moving users are attended by the macrocell layer while slow-moving users are attended by the microcell layer. Overflow traffic, and hand-ups and hand-downs are considered. The DDRA strategy is implemented at both the carrier and time slot levels, and the benefits of BS synchronization are highlighted. The DRA proposed is compared with the FCA strategy and with an enhanced FCA version having overflow traffic, hand-downs, and time slot reallocations.

1.2.6   Use of DDRA in Realistic HCS with a UMTS MAS

To show how the techniques can be applied to a likely third generation system, DDRA is used with a two-layer HCS consisting of "Manhattan" microcells overlaid with hexagonal macrocells in Chapter 7. The MAS scheme used is a hybrid WB-TD/CDMA-FDD. Channel borrowing within cells is taken into account as well as hand-ups and hand-downs. The whole is assessed by computer simulation in terms of the proportion of call attempts blocked and numbers of calls forced to terminate early.

Finally, in Chapter 8, all the material of the book is drawn together and important features are highlighted for the future.

2

Resource Allocation Strategies—an Overview

2.1 What Is a CAS?

The name "channel allocation strategy" is given to the technique used to make the most efficient and equitable use of the available radio spectrum, by the way in which channels are allocated to calls, and channels to cells and hierarchical levels on a fixed or dynamic basis.

2.1.1 Disjoint Channels

The first step in any WCMTS is to divide the spectrum allocation into a set of disjoint noninterfering radio channels so that they can be used simultaneously in an area of service, while maintaining an acceptable received radio signal in each [1]. Then, using a suitable MAS, physically scattered users can share a common uplink (UL). This is a very important design issue for efficient and fair utilization of the available system resources. A good MAS can improve system capacity, lower system cost, and make the service more attractive to users in terms of fewer dropped calls and interruptions [2]. Simpler types of MAS—such as time division multiple access (TDMA), frequency division multiple access (FDMA), and CDMA—merely use the total available spectrum to form such channels in ways with which the reader will be familiar. From these, more complex techniques consisting of a combination of two or more MASs may be used to generate a set of disjoint channels. Some of these channels are used to perform call set-up functions, and the rest are used to carry the offered traffic.

2.1.2 Cellular Reuse of Channels

A key feature of cellular systems is the reuse of identical channels in cells far enough apart so that the signals do not interfere. How the channels are to be assigned for simultaneous use in different cells directly affects the throughput of such systems [3]. The quality of the received signal that can be achieved in each channel and the cochannel interference achieved by frequency reuse are the most important factors in determining the number of channels with a certain quality that can be used for a given wireless spectrum. They are also the most serious restrictions on the overall system capacity in wireless networks.

The main idea behind a CAS is to improve the CIR using the radio propagation path loss characteristics between cells that use identical channels. Hence, a CAS is defined as the management and administration of the available radio channels in a WCMTS, that is, assigning them for simultaneous use in the different cells that form part of the system in such a way that the CIR is maximized. Ideally, the basis for channel assignment should be to maintain the least interference in the system; but, unfortunately, most cellular systems cannot be optimized in this way [4] due to such factors as terrain irregularities and man-made obstacles. The performance of a CAS should also result in an increase of the spectrum efficiency. In addition, a CAS must be flexible enough to handle dynamic system reconfiguration and nonuniform traffic. In general, there is a trade-off between performance improvements and system complexity [2].

2.1.3 Types of CASs and Their Classification

There are many features that can be used to classify CAS techniques. Each has its particular features and applications.

The most common basis is to compare the CASs in terms of the manner in which cochannels cells are separated. In this classification, CASs are divided into FCA, DCA, and, combining the first two, HCA strategies. These are discussed further later in this chapter.

A second important classification of CASs is based on the way they are implemented. They can either be centralized or distributed. When using centralized implementation, channels are assigned by a central controller (i.e., BSC); whereas with distributed implementation, channels can be selected either by the mobile in an autonomous way or by the local BS after performing measurements on the set of possible channels available. This results in better spectrum efficiency for centralized rather than distributed CASs, but a much greater signaling load results because of the huge amount required between BSCs and the switches in order to know the status of every channel at the time of allocation. Distributed CASs require much less signaling because each BS in each cell keeps information about the status of current available channels in its neighborhood and every change is communicated only between the BSs involved. In truly autonomously organized distributed schemes, the mobile chooses a channel based on its local CIR measurements without the involvement of a central call assignment entity at all [5, 6]. Obviously, most often minimized complexity in the implementation of CASs results in a lower realized efficiency.

CIR measurements can also be used as the basis for channel assignment. This can be done centrally, assigning channels to cells so that the minimum CIR requirement is fulfilled with high probability in all the radio cells involved, or distributed, using CIR values measured by MSs or local BSs [5, 6].

Handover techniques are of special importance in any cellular system because there is nothing more irritating to a customer than to lose his call when moving from one cell to another. It is usually preferable to block a new incoming call than to lose an existing one. Several techniques are used to implement handover in ways that cope well with traffic variation while maintaining a high level of utilization. Many other factors and techniques are used in particular circumstances and several are examined later in this chapter.

Finally, and of great importance for the future, particular techniques are needed for efficient channel utilization in HCSs, which have more than one cell layer. They introduce the extra degree of freedom of hand-up and hand-down. HCSs are of special importance to the work in this book.

The rest of the chapter summarizes particular techniques.

2.2 FCA

A description of the classic FCA strategy is given in this section. In addition, some strategies that are based on FCA but have additional features that allow them to perform better are also described.

2.2.1 Reuse Distance and Cell Patterns

In the FCA strategy, the area of service is partitioned into a number of cells and the same number of channels is permanently assigned to each cell according to a reuse pattern [7]. These channels are called nominal channels. The same set of channels can be reused in another cell provided that the reuse distance D is fulfilled [7]. The reuse distance is the minimum separation of identical channels that have the same carrier frequency, at which there is acceptable interference (CIR)

$$D = \sqrt{3N} \cdot R \tag{2.1}$$

where R is the cell radii and N is the reuse pattern. So if we consider the total number of channels available in the system to be M, the set of channels given to each cell, S, is easily found by

$$S = \frac{M}{N} \tag{2.2}$$

with

$$N = \frac{1}{3}\left(\frac{D}{R}\right)^2 \tag{2.3}$$

N can assume only integer values 1, 3, 4, 7, 9, 12, ... as generally presented by the series $(i + j)^2 - ij$, where i and j are integers [7–9]. It is obvious that the shorter the channel reuse distance, the greater the channel reuse over the whole service area. Figures 2.1(a, b) show the allocation of channel sets to cells for $N = 3$ and $N = 7$ in an environment formed of hexagonal shaped cells.

FCA strategy is very simple and easy to implement. The uniform distribution of channels performs very well under uniform traffic loads (where the same

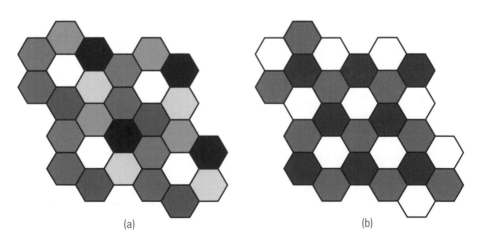

(a)   (b)

Figure 2.1 Examples of reuse patterns: (a) $N = 7$ and (b) $N = 3$.

traffic load is offered to every cell in the system) because the same number of channels is given to every cell. In FCA, a new call can only be served by the nominally assigned channels. If all the channels are busy, then the call is blocked from the system. In the case when the traffic is uniform—that is, the same offered load per cell—the overall probability that a new call is blocked from the system is equal to the probability that a user is blocked from a particular cell because every cell has the same number of channels and is being offered the same offered load.

The FCA strategy is a good option with which to start when a new system is deployed and the offered load is anticipated to be even. It also performs well under heavy traffic loads. However, the traffic distribution in mobile systems is more often nonuniform with spatial and temporal fluctuations, therefore, a nonuniform allocation of channels in every cell is required to match the load offered to them. One of the main drawbacks of FCA when it has to face nonuniform traffic distributions varying in both time and space is that a uniform allocation of channels may result in a high blocking probability in cells with high offered loads, and a low blocking probability in cells with low offered loads. This is mainly because simple FCA cannot reallocate channels. FCA clearly lacks the flexibility to share the resources between adjacent cells to cope with varying traffic conditions. The end of a movie in a cinema or the end of a play in a theater may lead to such a situation when several people use their mobile telephones to call someone to relate how good the performance was. This sudden demand of service in a cell covering this hot spot might create high blocking probability in it, whereas neighboring cells might have idle channels and experience low blocking probabilities. There are several situations that might produce similar scenarios. This is clearly unacceptable.

One way to relieve this problem is to use FCA strategies with nonuniform channel allocation whereby the number of nominal channels assigned to each cell depends on the expected load submitted to the cell, based on past experience or statistics. Thus, heavily loaded cells are given more channels than lightly loaded cells [10, 11], improving the channel utilization within the cells. This allocation of channels will be more adequate for situations like those mentioned previously; however, it still will not have attained an optimum performance level.

2.2.2 Channel Borrowing Schemes

Another way to overcome the effects of nonuniform loading is to borrow free channels from neighboring cells.

In strict technical terms this is described by saying that when there are no free nominal channels in a cell, a channel from a neighboring cell may be borrowed to accommodate new calls if the interference constraints are fulfilled. Channels are borrowed on the basis that they will cause the least harm in the neighboring cells [12]. When a channel is borrowed, several other cells are prohibited from using it. This is called *channel locking* [3]. The number of cells in which the borrowed channel has to be locked depends on the reuse pattern N being used, the type of cell layout, and the type of initial allocation of channels to cells. For example, considering Figure 2.2, if a reuse patter of 7 is used and a channel borrowing has occurred (cell P has borrowed a channel from cell A_3), then the channel has to be locked in the first and second tiers of cells surrounding the cell that borrowed the channel (cell P)—this includes cochannel cells A_2 and A_4.

Channel borrowings are temporary and last only for the duration of a call. Once calls are completed, the borrowed channels are returned to the their original cells and locked channels are released. The channel borrowing scheme is used primarily for slowly growing systems. It is often helpful in delaying cellsplitting [7] in peak traffic areas. Since cell splitting is costly, it should be implemented only as a last resort [4].

Channel borrowing strategies only need local and neighboring cell information. Several channel borrowing strategies have been proposed in the literature [13–28]. They can be classified into *simple* and *hybrid* [3, 16]. They differ in the way a free channel is selected from a neighboring cell.

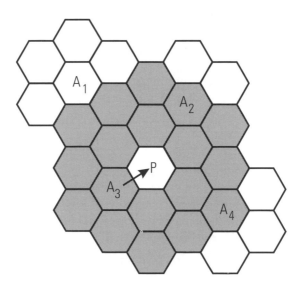

Figure 2.2 Channel borrowing and directional locking.

Channel borrowing schemes perform better than FCA under low and moderate traffic loads, but fail to perform better than FCA under heavy loads, because when traffic is low, the number of borrowed channels is small and they cope with the fluctuations of the offered traffic, but when the offered load is high, the number of borrowings may proliferate to such an extent that the channel utilization drops drastically [1]. This is caused by channel locking. It may happen, nevertheless, that the set of neighboring cells from which a channel can be borrowed contain several channels available to lend. In such a case, an algorithm is used to select one of the candidate channels for borrowing. This algorithm is called the *cost function*.

2.2.3 Simple Channel Borrowing

In simple channel borrowing any nominal channel from a neighboring cell can be borrowed at any time as long as the channel is free and the borrowing does not cause interference with calls already in progress in neighboring cells. Several simple channel borrowing strategies have been proposed [1, 13].

- Borrow from the richest (SBR);
- Basic algorithm (BA);
- Basic algorithm with reassignment (BAR);
- Borrow first available (BFA).

In SBR every idle channel in a neighboring cell is a candidate to be borrowed. If, however, more than one neighboring cell has idle channels available, a channel is borrowed from the cell with the most available idle channels. However, SBR does not take into account channel locking when choosing a candidate channel for borrowing [1]. The BAR strategy improves upon SBR by taking into account channel locking when selecting a channel for borrowing. BAR tries to minimize the future call blocking probability in the cell that is most affected by the channel borrowing by choosing the channel that maximizes the number of available nominal channels in the worst case nominal cell within the reuse distance. BAR makes use of intracell channel reallocations to transfer a call from a borrowed channel to a nominal channel whenever a nominal channel becomes available. The borrowed channel to be released is the channel that will minimize the blocking probability of the donor cell. The BFA strategy selects the first candidate channel it finds without attempting to optimize when borrowing. Here the philosophy of the nominal channel assignment is also different. Instead of assigning channels directly to cells, the channels are subdivided into sets and then each set is assigned to cells at a reuse distance. These sets are numbered in sequence. When setting up a call, channels are searched in a prescribed way to find a candidate channel [1].

2.2.4. Hybrid Channel Borrowing

In hybrid channel borrowing, the nominal set of channels in every cell is subdivided into two subsets of channels, A (containing nominal channels that can only be used by the owner cell) and B (containing channels that can be lent to neighboring cells). For a fixed ratio A/B, the number of possible lending cells increases with larger N but the number of candidate channels per cell decreases. By incorporating channel reassignment into these strategies, a maximum packing of channels can be obtained, hence increasing channel utilization [3, 21, 23–26].

There are also several examples of hybrid channel borrowing schemes, such as:

- Simple hybrid channel borrowing (SHCB) [3];
- Borrowing with channel ordering (BCO) [3, 16];
- Borrowing with directional channel locking (BDCL) [3];
- Channel assignment with borrowing and reassignment (CARB) [14];
- Ordered channel assignment scheme with rearrangement (ODCA) [21].

In SHCB, the ratio of $|A|$ to $|B|$ (the number of nominal channels and the number of borrowable channels, respectively) is determined a priori based on traffic measurements or assessments and can be adapted dynamically in a scheduled or predictive manner. BCO is an improvement of SHCB that allows the variation of the nominal-to-borrowable channel ratio according to traffic conditions. In BCO nominal channels are ordered so that the first channel has the highest priority to attend the next local call and the last channel is given the highest priority for being borrowed by the neighboring cells. BCO [3] also allows channel reallocations to release borrowed channels as soon as there is a nominal channel available. In BCO, a channel borrowing can only be made if a channel is simultaneously free in the three nearby cochannel cells. In BDCL, the channel locking in the cochannel cells is restricted to those directions affected by the borrowing only, therefore increasing the number of channels to be borrowed. As in BCO, channel reallocations are used in BDCL to transfer calls on borrowed channels to nominal channels and also between borrowed channels. In CARB channels are borrowed on the basis of causing the least harm to neighboring cells in terms of future blocking probability. Reassignment of borrowed channels is done so as to cause maximum relief to neighboring cells. ODCA is a combination of CARB and BCO that produces higher performance.

2.3 DCA

DCA overcomes the main drawback of FCA strategies—no or little adaptability to temporal and spatial variations of traffic—by assigning channels in a dynamic way. In contrast to FCA, in DCA strategies there is no relationship between the channels and the cells and every channel can be used in any cell as long as the cochannel reuse distance and the interference constraints are fulfilled [27–61]. The channels are used in the cells only for the duration of the call [3]. Arrivals are attended on a call-by-call basis, searching for the channel with minimum cost to the system. Normally, in this strategy there exists a central pool in charge of assigning channels to each cell according to a cost function that maximizes channel efficiency (whenever possible). As soon as the calls are over, the channels are returned to the central pool. The main idea of all DCA strategies is to evaluate the use of a candidate channel by means of the cost function and to select those with minimum cost to the system. Different DCA schemes are distinguished by the type of the cost function [22]. This can be a function of the future expected load [23], the future overall or average blocking in the system [24], the usage frequency of the candidate cell, the reuse distance, channel occupancy distribution under current traffic conditions, or radio channel measurements of individual mobile users [25]. In practice, DCA strategies are divided into three main groups that exploit one of the following characteristics [29]:

- Adaptability to interference;
- Adaptability to traffic variations;
- Channel reusability [29].

DCA algorithms can adapt automatically to varying traffic conditions in space and time [30]; there is no need for information on the channel status during the assignment procedure in order to determine whether or not it is possible to use some channel in a specific case. DCA algorithms do not require channel planning and have shown a better performance than FCA strategies under low and moderate traffic load conditions [3]. However, when heavy traffic load conditions are present in the system, DCA strategies do not outperform FCA strategies, especially when the traffic is nonuniform. This behavior occurs because cells that are given the same set of channels are separated, on average, by a distance larger than the minimum cochannel reuse distance. To improve this situation, channel reassignment can be used to pack the cochannel cell closer together. Channel reassignment means, wherever possible, switching calls in progress to other channels to reduce the distance between cochannel cells [24]. Thus, the channel reuse distance is shortened, and more traffic can be carried per channel at a given blocking rate. There is actually a limit to the packing of resources. This limit is given by the *maximum packing* strategy [1], which limits the performance of every DCA strategy that fulfills the reuse distance.

The maximum packing DCA policy is a good idealization of DCA that allows analytic tractability. Specifically, maximum packing assumes that a call will be blocked if there is no possible reallocation of channels to calls (including reallocation of calls in progress) that would result in the call being carried. The usefulness of this policy is that it provides a good indication of the performance of DCA algorithms without going into the details of specific algorithms.

Perhaps the major disadvantage that DCA strategies have in comparison with FCA strategies is the fact that not only the transceivers for the nominal channels must be available at every BS but, in some cases, the transceivers of all of the channels. The number required depends on the DCA strategy itself. This drawback is manifested in economic terms rather than in functionality.

As in the case of FCA strategies, there are several classifications of DCA strategies based on different comparison bases. One way is to classifying them into *call-by-call* DCA and *adaptive* DCA schemes. In a call-by-call DCA scheme the assignment is based on the current usage conditions in the cell, whereas in an adaptive DCA scheme the channel assignment is adaptively carried out using information on the previous and present channel usage conditions [31, 57]. But perhaps the most important classification of DCA strategies is that which distinguishes them as *centralized* and *distributed*.

2.3.1 Centralized DCA Schemes

In centralized DCA strategies channels are assigned to incoming calls by a central controller from a central pool. These strategies differ from each other in thetype of cost function used to select the candidate channel to attend incoming calls [32–37]. The use of a central controller implies a huge amount of signaling load produced by communication between each BSC and the central controller (normally the switch) in order to know the status of the channels in every cell at the time when an incoming call requests a channel. Thus, the delay in allocating a channel in centralized DCA strategies is proportional to the number of BSs that cover the area of service. The huge amount of computational time required by most of the centralized DCA strategies helps to increase the allocation delay. That is why this type of DCA strategy is applied primarily to macrocellular environments [32, 35], where a large geographical area is served by cells of large radii. Obviously, this does not exclude its use in macrocellular environments [36, 37]. In fact, the use of centralized DCA strategies in this type of environment is reemerging [36, 37]. The availability of channel status information makes the efficiency of centralized DCA strategies far better than distributed DCA strategies because the knowledge of the channels status in every single cell facilitates making a better decision on which channel to use. Several simulation and analysis results have shown that centralized DCA schemes can produce near-optimum channel allocation at the expense of a high centralization overhead [41–43, 57].

There are several examples of centralized strategies proposed in the literature, such as:

- First available (FA) [22];
- Locally optimized dynamic assignment (LODA) [3, 10];
- Selection with maximum usage on the reuse ring (RING) [22].

The simplest of them is the FA strategy. In FA, the first available channel within the reuse distance encountered during a channel search is assigned to the call. Since no cost function is evaluated to select the optimum channel, the FA minimizes the system computational time. In the LODA strategy, the selected cost is based on the future blocking probability in the vicinity of the cell in which a call is initiated [1]. In the RING strategy, a candidate channel is selected that is in use in the most cells in the cochannel set. If more than one channel has this maximum usage, an arbitrary selection among such channels is made to serve the call. If none is available, the selection is made based on the FA scheme.

2.3.2 Distributed DCA Schemes

In microcellular systems, as explained previously, distributed DCA strategies [38–61] are more attractive for implementation than centralized DCA strategies due to the simplicity of the assignment algorithm in each BS. However, propagation characteristics will be less predictable and network control requirements more intense than in macrocellular systems [1]. Several simulation and analysis results have shown that centralized DCA schemes can produce near-optimum channel allocation but at the expense of a high centralization overhead. Distributed schemes are therefore more attractive for implementation in microcellular systems due to the simplicity of the assignment algorithm in each BS.

Distributed DCA strategies are less stable than centralized schemes because the former are affected by local changes (e.g., traffic variations) that could spread through the whole system, hence affecting its performance. Centralized systems, on the other hand, may detect local variations and cope with them quickly, avoiding abrupt fluctuations in the blocking probability in areas with traffic problems.

Distributed DCA strategies can be classified into those that rely on information about the channel status in neighboring cells (cell based) [44–47] and those that relay on signal strength measurements [48–50]. In the cell-based strategies, the BSs assign channels to incoming calls based on information about the current status of channels in their vicinity. This information is updated continuously. The performance of cell-based DCA strategies is not as optimum as that shown by centralized DCAs, and signaling load between BSs increases as the offered load increases.

In DCA strategies that assign channels relying purely on local signal strength measurements there is no need for a BS to communicate with any other BS in the network. Thus, the system is self-organizing and cellular planning is totally avoided. The delay in the channel assignment process is practically nil. These types of strategies allow maximum packing only at the expense of increasing cochannel interference in ongoing calls in adjacent cells. This may produce forced termination and deadlocks [48].

There have been a huge number of distributed DCA strategies proposed in the literature over the years and explaining them here would require more space than is available. Readers are encouraged to see the references.

2.3.2.1 One Dimensional DCA Strategies

Distributed DCA strategies can also be applicable to one-dimensional cellular mobile systems. One-dimensional structures can be identified in cases such as streets with tall buildings shielding interference on either side [29]. They can also be highways or motorways [21, 40, 46, 47, 51, 57].

The minimum interference (MI) scheme [29], the random minimum interference (RMI), the random minimum interference with reassignment(RMIR), the sequential minimum interference (SMI), and the MINMAX algorithms are some examples of distributed DCA strategies proposed for one-dimensional cellular systems.

The MI scheme is well known and among the simplest for one-dimensional cellular systems. It is incorporated in the enhanced cordless telephone (CT-2) and digital European cordless telephone (DECT) systems. In this scheme, a mobile signals its need for a channel to the nearest BS. The BS measures the interfering signal power on all channels not already assigned to other mobiles. The mobile is assigned the channel with minimum interference [1]. In the RMI scheme, the mobiles are served according to the MI scheme in a random order or, equivalently, the order in which calls arrive in the system. In RMIR, mobiles are also served according to the MI scheme, and each mobile is then reassigned a channel according to the MI scheme. Those mobiles denied service by the initial RMI scheme also try to obtain a channel again [1]. The order in which mobiles are reassigned is random. The number of times this procedure is carried out is the number of reassignments, R.

In the SMI scheme, mobiles are assigned channels according to the MI scheme but following a sequential order. The sequence is such that any mobile is served only after all the mobiles ahead of it have had a chance to be served. This process requires some coordination between BSs because of the sequential order of service [1]. Finally, in MINMAX a mobile is assigned a channel that maximizes the minimum of the CIRs of all mobiles being served by the system at that time. A mobile is served only after all mobiles to the left of it have had a chance to be served. This sequential order (left to right) of service is chosen because it appears to be the best way to reuse the channel [29]. The mobile immediately to the right of a given set of mobiles with channels assigned is the one that will cause the most interference at the BS servicing the given set of mobiles and is also that which has the most interference from that set of mobiles.

2.4 HCA

In HCA strategies the total set of channels is divided into two subsets. The first subset of channels is assigned to the cells of the system according to the FCA strategy. The second subset is kept in a central pool and assigned dynamically to the cells on demand to increase flexibility. Therefore, there are basically two types of CAS at the same time, FCA and DCA. When a new call arrives at a cell, the first attempt to serve it is by a nominal or fixed channel. If there is no free nominal channel, a channel from the dynamic set is assigned to the call. If this fails, then the call is blocked. The way in which the dynamic channels are managed follows any of the strategies in [32–61]. The ratio of fixed-to-dynamic channels is a significant parameter that defines the performance of the system in much the same manner that the ratio of nominal-to-borrowable channels defines the performance of a strategy with channel borrowing. In general, the ratio of fixed-to-dynamic channels is a function of the traffic load and would vary over time according to offered load distribution estimations [1].

2.5 Other Strategies

There are several schemes that do not fit into the classifications in the previous sections. They are described in the following subsections.

2.5.1 CAS Handling Handovers

A handover is defined as a change of radio channel used by a mobile terminal. The new channel may be within the same cell (intracell handover) or in a different cell (intercell handover). A handover can be initiated by channel real-location or by link degradation. They are an important issue in microcellular systems, where the cell radius is small, because the performance of such systems depends significantly on the number of handovers performed by mobile users. Of particular interest in these systems is the reduction of calls forced to terminate, because from both the user and the service provider point of view, it is less desirable to force an ongoing call to terminate than to block a new call. Therefore, methods for decreasing the probability of forced termination by prioritizing handovers at the expense of a tolerable increase in call blocking probability have been devised in order to increase the quality of cellular service [18, 62–69]. These methods are the NPS [63], the RCS [62, 63], the first in first out (FIFO) priority scheme [67], the measured based priority scheme [67], and the subrating scheme (SRS). They are classified into three categories: basic, queuing [66], and subrating [64].

There are no restrictions on combining any of these channel allocation strategies for handover with any of the FCA, DCA, or HCA strategies described previously in order to tackle both the blocking of new users and the blocking of handover calls.

2.5.1.1 NPS

In the NPS, BSs handle the handovers in exactly the same manner as new call arrivals; therefore, the blocking of handover calls is the same as the blocking of new calls.

2.5.1.2 Prioritized Scheme

The simplest way to give priority to handover calls is by specially reserving channels for them in every cell. The guard concept was introduced in [62, 63]. In this scheme the available channels in a cell are divided into two sets: A (a set of channels that attend nominal new calls as well as handover calls) and B (a set of channels that attend handover calls only). This scheme provides improved performance at the expense of a reduction in the total admitted traffic and an increase in the blocking of new calls. Another shortcoming of the employment of guard channels, especially with fixed channel assignment strategies, is the risk of inefficient spectrum utilization. Allowing the queuing of new calls may ameliorate this disadvantage.

The queuing of handover requests is another generic prioritization scheme offering reduced probability of forced termination [65, 66]. Queuing handover techniques can be used in conjunction with guard channels. In this strategy there is again a trade-off with the increase in total carried load [18]. The scheme is described as follows [1]. When the power level received by the BS in the current cell falls to a certain threshold, namely the handover threshold, the call is queued for service from a neighboring cell. The call remains queued until either an available channel in the new cell is found or the power by the BS in the current cell drops below a second threshold called the receiver threshold. If the call reaches the receiver threshold and a new channel has not been found, then the call is terminated. Queuing handover requests is made possible by the existence of the time interval that the MS spends between these two thresholds. This interval defines the maximum allowable waiting time in the queue. Based on the traffic pattern and the expected number of requests, the maximum size of the handover queue can be determined [1]. A handover may still be dropped because the handover request has no choice but to wait until the receiver threshold is reached; so when there is a high demand for handovers, they will be denied queuing due to the limited size of the handover queue. The basic queuing discipline in a queuing handover request is FIFO [18].

2.5.1.3 SRS

In the SRS [64], if a BS does not have a free channel to attend a handover call, a new channel is created to attend it by subrating an existing call. Subrating means an occupied full rate channel is temporally divided into two channels at half the original rate: one serves the original call and the other serves the handover request [64, 69]. The blocking probabilities (combined forced termination of existing calls and blocking of new call attempts) of this new scheme compare favorably with the standard scheme (nonprioritizing) and the schemes proposed previously. However, this scheme presents an additional complexity of implementing on-the-fly subrating and the impact of continuing the conversation on a lower rate channel (which may lower speech quality or increase battery drain) [64].

2.5.2 Reuse Partitioning

In reuse partitioning each cell is divided into two or more concentric zones or subcells [70–84]. The reuse distance is smaller for the inner cells than for the outer cells. The smaller the zone, the smaller the reuse pattern used (obviously, the smallest reuse pattern an inner cell can make use of is 1). This results in high spectrum efficiency.

The reasoning behind the reuse partitioning strategy is that, because the inner zones are closer to the BS located at the center of the cell, the power level required to achieve a desired CIR in the inner zones can be much lower than in the outer zones. This can be appreciated in Figure 2.3. In reuse partitioning, several overlaid fixed channel plans are used. Mobiles with favorable positions and high signal levels are assigned channels from plans with low reuse distances, whereas mobiles with low signal quality get channels with large cluster sizes [79].

Reuse partitioning strategies can be divided into fixed [70–73] and adaptive strategies [74–84]. Basically, in fixed reuse partitioning, each specific zone (or subcell) is given a subset of nominal channels allocated according to the FCA strategy. This subset of channels cannot be changed. In adaptive channel allocation techniques combined with reuse partitioning schemes, any channel in the system can be used by any BS as long as the required CIR is maintained. It should be noted that reducing the CIR margin in each channel leads to an improvement in the traffic handling capacity.

2.5.3   Directed Retry

Directed retry (DR) [85, 86] and directed handover (DH) take advantage of the fact that some percentage of the mobile stations may be able to obtain sufficient quality from two or more cells because of overlapping coverage between adjacent cells. In DR, if a new call arrival finds that its first attempted cell has

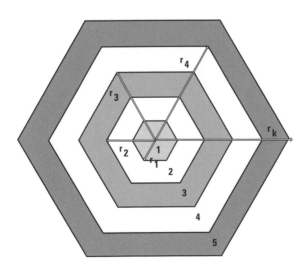

Figure 2.3 Reuse partitioning in hexagonal cells.

no free channels available, it will try to get a channel in any other cell that provides sufficient signal quality. The DH scheme takes this idea further, in that when a cell has all or almost all of its channels in use, it may, using DH, direct some of the calls currently in progress in its domain to attempt handover to an adjacent cell. The motivation here is to distribute calls from heavily loaded cells to lightly loaded cells [53].

2.5.4 The Bunch Concept

A bunch is defined as a group of remote antenna units (RAUs) strongly interconnected. It was first presented in [58, 87]. All resources available to a *bunch* are dynamically allocated by a radio network controller (RNC) [37]. A RAU can be a classical BS or a simple antenna. In this case, the RNC-RAU link can be an optical cable carrying the radio-frequency signal and the real transceivers may be located at the RNC level [88]. Between RAUs within a bunch, high-speed signaling can be used for channel allocation. The RAUs in a bunch are assumed to be synchronized. The RNC has complete knowledge of all allocated resources, including actual transmission powers for any channel at any time. The centralization at the RNC level permits the use of near-optimal algorithms for resource allocation and management (e.g., power control and DCA). For fast DCA, algorithms could range from the very simple RCA to complex ones such as the maximum packing algorithm (theoretical upper bound for DCA performance) [88]. The bunch size can be adapted to the type of environment. In the indoor case, a bunch can be a wireless PBX, for example. In the rural environment it may only include several macrocells [88].

Different techniques of resource management between bunches can be used—for instance, interference diversity strategies or segregation strategies. Interference diversity can be provided by frequency or time slot hopping desirable in particular at border zones. For example, the hopping sequence can be the same for all users in a bunch but different hopping sequences are assigned to neighboring bunches. In such a configuration, the RNCs do not need tocommunicate with each other. If a distributed process between RNCs for a slow allocation of frequencies to bunches is considered, low-speed signaling is then required [88].

2.5.5 Channel Allocation for HCS

The channel assignment strategies proposed for HCS are different from the schemes discussed in previous sections. As stated in Chapter 1, the most common HCS consists of two layers only: microcellular and macrocellular. Channel allocation strategies for HCS are motivated mainly to decrease the number of handover attempts in the microcellular layer and to increase channel utilization in the macrocellular layer [89–116]. Here, a cluster of microcells are grouped together and covered by a macrocell that works as an overflow group of servers. The total bandwidth is divided between microcells and macrocells in its domain. In case of congestion, if there are not enough microcell channels for handovers or new call arrivals, then the macrocell channels can be used. Because the macrocell BS covers a much larger area than a microcell, its transmitted power is higher than that of microcells [116]. That creates the need to use different allocation techniques to manage the resources in the layers and to optimize them. An important issue in these strategies is the point of sharing the channels not only between cells of the same hierarchy, but between cells of different hierarchies.

2.6 Summary

In WCMTS, CASs play an important role in increasing spectrum efficiency. For a given spectrum and a specific technology, the traffic-carrying capacity of a cellular system depends on how the radio channels are managed.

There are several classifications of CASs according to different comparison bases, but in the most widely accepted classifications they can be classified as FCA, DCA, and HCA. Here, they are classified according to the way in which the cochannel sets are separated. Conventionally, FCA is used where each cell is assigned a set of fixed nominal channels. If a new call arrival finds all the nominal channels busy, then the call is blocked from the system. In contrast, in DCA, there is no relationship between the channels and the cells in the system. Channels are given to the cells on demand only for the duration of the calls; as soon as they are finished, the channels are returned to a central pool that manages the resources. Any cell can use any channel as long as the reuse distance is fulfilled. An important feature of DCA strategies is their adaptability to varying traffic conditions. They can adapt to both time and space variations of traffic.

In the third category of this classification, the total available channels are divided into two subsets: one set of channels is assigned to each cell on a nominal basis and the others are kept in a central pool for dynamic use.

As the demand for WCMTS grows, the deployment of HCS is inevitable and cells will become smaller in metropolitan areas, leading eventually to a microcellular layer. This layer will have to be overlaid by a macrocellular layer. DCA will, therefore, play an important role in WCMTS with HCS to manage the resources in and between layers.

References

[1] Katzela, I., and M. Naghshineh, "Channel Assignment Schemes for Cellular Mobile Telecommunication Systems: A Comprehensive Survey," *IEEE Personal Commun. Mag.*, Vol. 3, No. 3, June 1996, pp. 10–31.

[2] Victor, O. K. L., and Xiaoxin Qiu, "Personal Communication Systems (PCS)," *Proc. IEEE*, Vol. 83, No. 9, Sept. 1995, pp. 1210–1243.

[3] Zhang, Ming, and Tak-Shing P. Yum, "Comparisons of Channel Assignment Strategies in Cellular Mobile Telephone Systems," *IEEE Trans. Vehicular Tech.*, Vol. 38, No. 4, Nov. 1989, pp. 211–215.

[4] Lee, William C. Y., *Mobile Cellular Telecommunication Systems*, London: McGraw-Hill, 1989.

[5] Zander, J., "Asymptotic Bounds on the Performance of a Class of Dynamic Channel Assignment Algorithms," *IEEE J. Select. Areas Commun.*, Vol. 11, No. 6, Aug. 1993, pp. 926–933.

[6] Chuang, J. C.-I., "Performance Issues and Algorithms for Dynamic Channel Assignment," *IEEE J. Select. Areas Commun.*, Vol. 11, No. 6, Aug. 1993, pp. 955–963,.

[7] MacDonald, V. H., "The Cellular Concept," *Bell Systems Techn. J.*, Vol. 58, Jan. 1979, pp. 15–41.

[8] Kahwa, T. J., and N. Georganas, "A Hybrid Channel Assignment Scheme in Large Scale Cellular-Structured Mobile Communication Systems," *IEEE Trans. Commun.*, Vol. COM 26, No. 4, April 1978, pp. 432–438.

[9] Lee, W. Y. C., *Mobile Cellular Communication Systems*, London: McGraw-Hill, 1989.

[10] Zang, M., and T. S. Yum, "The Nonuniform Compact Pattern Allocation Algorithm for Cellular Mobile Systems," *IEEE Trans. Vehicular Tech.*, Vol. 40, No.2, May 1991, pp. 387–391.

[11] Oh, S.-H., et al., "Prioritised Channel Assignment in a Cellular Radio Network," *IEEE Trans. Commun.*, Vol. 40, No. 7, July 1992, pp. 1259–1269.

[12] Ortigoza-Guerrero, Lauro, and Domingo Lara-Rodríguez, "CPMCB: A Suitable DCA Scheme for the Pan-European GSM System," in *Proc. IEEE Veh. Tech. Conf. VTC'97*, pp. 532–536, Phoenix, Arizona, May 1997.

[13] Anderson, L., "A Simulation Study of Some Dynamic Channel Assignment Algorithms in High Capacity Mobile Telecommunication Systems," *IEEE Trans. Vehicular Tech.*, Vol. VT-22, No. 4, Nov. 1973, pp. 210–217.

[14] Engel, J. S., and M. Peritsky, "Statistically Optimum Dynamic Server Assignment in Systems with Interfering Servers," *IEEE Trans. Vehicular Tech.*, Vol. VT-22, No. 4, Nov. 1973, pp. 203–209.

[15] Sawada, H., et al., "Techniques for Increasing Frequency Spectrum Utilization," IECE Technical Report CS84-100, 1984.

[16] Elnoubi, S. M., R. Singh, and S. C. Gupta, "A New Frequency Channel Assignment Algorithm in High Capacity Mobile Communication Systems," *IEEE Trans. Vehicular Tech.*, Vol. VT-31, No. 3, Aug. 1982, pp. 125–131.

[17] Johri, P., "An Insight into Dynamic Channel Assignment in Cellular Mobile Communication Systems," *Euro J. Operational Research*, Vol. 74, 1994, pp. 70–77.

[18] Tekinay, S., and B. Jabbari, "Handover and Channel Assignment in Mobile Cellular Networks," *IEEE Commun. Mag.*, Vol. 29, No. 11, Nov. 1991, pp. 42–46.

[19] Tum, T. S. P., and M. Schwartz, "The Join-Biased-Queue Rule and its Applications on Routing in Computer Communication Networks," *IEEE Trans. Commun.*, Vol. COM-29, No. 4, April 1981, pp. 505–511.

[20] Yum, T. S. P., and W.-S. Wong, "Hot Spot Traffic Relief in Cellular Systems," *IEEE J. Select. Areas Commun.*, Vol. 11, No. 6, Aug. 1993, pp. 934–940.

[21] Kuek, S. S., "Ordered Dynamic Channel Assignment Scheme with Reassignments in Highway Microcell," *IEEE Trans. Vehicular Tech.*, Vol. 41, No. 3, Aug. 1992, pp. 271–277.

[22] Cox, D. C., and D. O. Reudink, "Dynamic Channel Assignment in Two Dimension Large Scale Mobile Radio Systems," *Bell System Tech J.*, Vol. 51, Sept. 1972, pp. 1622–1628.

[23] Ortigoza-Guerrero, Lauro, and Domingo Lara-Rodríguez, "A Compact Pattern with Maximised Channel Borrowing Strategy for Mobile Cellular Networks," in *Proc. IEEE PIMRC'96*, pp. 329–333, Taipei, Taiwan, Oct. 1996.

[24] Lawrence Yeung, Kwan, and Tak-Shing Peter Yum, "Compact Pattern Based Dynamic Channel Assignment for Cellular Mobile System," *IEEE Trans. Vehicular Tech.*, Vol. 43, No. 4, Nov. 1994, pp. 892–896.

[25] Maric, S. V., et al., "Adaptive Borrowing of Ordered Resources for the Pan-European Mobile Communication (GSM) System," *IEE Proc. Commun.*, Vol. 141, No. 2, April 1994, pp. 93–97.

[26] Jiang, Hua, and Stephen S. Rappaport, "CBWL: A New Channel Assignment and Sharing Method for Cellular Communication Systems," *IEEE Trans. Vehicular Tech.*, Vol. 43, No. 2, May 1994, pp. 313–322.

[27] Chang, Kun-Nyeong, et al., "An Efficient Borrowing Channel Assignment Scheme for Cellular Mobile Systems," *IEEE Trans. Vehicular Tech.*, Vol. 47, No. 2, May 1995, pp. 602–608.

[28] Noubir, Guevara, "Inter-Layer Resource Management for Hierarchical Cell Structures," in *Proc. PIMRC'98*, Boston, MA, USA, Sept. 1998.

[29] Goodman, David J., Sudheer A. Grandhi, and Rajiv Vijayan, "Distributed Dynamic Channel Assignment Schemes," in *Proc. IEEE Veh. Tech. Conf. VTC'93*, pp. 532–535, Secaucus, New Jersey, USA, May 18–20, 1993.

[30] Yeung, K. L., and T. S. P. Yum, "The Optimisation of Nominal Channel Allocation in Cellular Mobile Systems," in *Proc. IEEE Int. Conf. on Commun. ICC'93*, Geneva Switzerland, pp. 915–919, May 1993.

[31] Okada, K., and F. Kubota, "On Dynamic Channel Assignment Strategies in Cellular Mobile Radio Systems," *IEICE Trans. Fundamentals*, Vol. 75, 1991, pp. 1634–1641.

[32] Cox, D., and D. Reudink, "A Comparison of Some Channel Assignment Strategies in Large Mobile Communication Systems," *IEEE Trans. on Commun.*, Vol. 20, No. 2, April 1972, pp. 190–195.

[33] Okada, K., and F. Kubota, "On Dynamic Channel Assignment in Cellular Mobile Radio Systems," in *Proc. IEEE Int. Symposium on Circuits and Systems*, Vol. 2, 1991, pp. 938–941.

[34] Goodman, D., J. Grandhi, and A. Sudheen, "Distributed Channel Assignment Schemes," in *Proc. IEEE Vehicular Technol. Conf. VTC'93*, pp. 532–535, Secaucus, New Jersey, USA, May 18–20, 1993.

[35] Del Re, E.,et al., "Handover and Dynamic Channel Allocation Techniques in Mobile Cellular Networks," *IEEE Trans. Vehicular Tech.*, Vol. 44, No. 2, May 1995, pp. 229–237.

[36] Iera, A.,et al., "Centralised Dynamic Resource Assignment in Microcellular Environments with Variable Load Distribution," in *Proc. IEEE Int. Conf. on Commun. ICC'97*, pp. 371–374, Montreal, Canada, June 8–12, 1997.

[37] Mihailescu, C., et al., "Locally Centralized Dynamic Resource Allocation Algorithm for the UMTS in Manhattan Environment," in *Proc. PIMRC'98*, Boston, MA, USA, Sept. 1998.

[38] Mcfarlane, D. A., and S. T. S. Chia, "Micro-Cellular Mobile Radio Systems," *BT Tech. J.*, Vol 8, No. 1, Jan. 1990, pp. 79–84.

[39] Rustakoet, A. J., et al., "Radio Propagation Measurements at Microwave Frequencies for Microcellular Mobile Communications," *IEEE Trans. Vehicular Tech.*, Vol. 2, 1991, pp. 203–210.

[40] Frodigh, M., "Reuse Partitioning Combined with Traffic Adaptive Channel Assignment for Highway Microcellular Radio Systems," in *Proc. IEEE GLOBECOM'92*, pp. 1414–1418, Orlando, Florida, USA, Dec. 1992.

[41] Gamst, A., "Some Lower Bounds for a Class of Frequency Assignment Problems," *IEEE Trans. Vehicular Tech.*, Vol. 35, No. 1, Feb. 1986, pp. 8–14.

[42] Sallberget, K., et al., "Hybrid Channel Assignments and Reuse Partitioning in a Cellular Mobile Telephone System," in *Proc. IEEE Vehicular Technol. Conf. VTC'87*, pp. 405–411, 1987.

[43] Nettleton, R. W., "A High Capacity Assignment and Reuse Partitioning in a Cellular Mobile Telephone System," in *Proc. IEEE Vehicular Technol. Conf. VTC'89*, pp. 359–367, 1989.

[44] C.-L. I. and P.-H. Chao, "Local Packing–Distributed Dynamic Channel Allocation at Cellular Base Station," in *Proc. IEEE GLOBECOM'93*, pp. 293–301, Houston, TX, USA, Nov. 29–Dec. 2, 1993.

[45] C.-L. I. and P.-H. Chao, "Distributed Dynamic Channel Allocation Algorithms with Adjacent Channel Constraints," in *Proc. IEEE PIMRC'94*, pp. 169–175, 1994.

[46] Okada, K., and F. Kubota, "Performance of a Dynamic Channel Assignment Algorithm with Information of Moving Direction in Mobile Communication Systems," in *Proc. IEICE Spring National Conf.*, pp. 334–338, 1991.

[47] Okada, K., and F. Kubota, "A Proposal of a Dynamic Channel Assignment Strategy with information of Moving Directions," *IEICE Trans. Fundamentals*, Vol. E75-a, 1992, pp. 1667–1673.

[48] Serizawa, M., and D. Goodman, "Instability and Deadlock of Distributed Dynamic Channel Allocations," in *Proc. IEEE Vehicular Technol. Conf. VTC'93*, pp. 528–531, Secaucus, New Jersey, USA, May 1993.

[49] Furuya, Y., and Y. Akaiwa, "Channel Segregation–A Self Organized Dynamic Allocation Scheme for Mobile Communication Systems," *IEICE Trans.*, Vol. 74, 1991, pp. 1531–1537.

[50] Akaiwa, Y., and H. Andoh, "Channel Segregation–A Self Organized Dynamic Allocation Method: Application to TDMA/FDMA Microcellular System," *IEEE J. Select. Areas Commun.*, Vol. 11, No. 6, Aug. 1993, pp. 949–954.

[51] El-Dolil, S. A., W. C. Wong, and R. Steele, "Teletraffic Performance of Highway Microcells with Overlay Macrocell," *IEEE J. Select. Areas Commun.*, Vol. 7, No. 1, Jan. 1989, pp. 71–78.

[52] Biaocchi, Andrea, et al., "The Geometric Dynamic Channel Allocation as a Practical Strategy in Mobile Networks with Bursty User Mobility," *IEEE Trans. Vehicular Tech.*, Vol. 44, No. 1, Feb. 1995, pp. 14–23.

[53] Everitt, David, and David Manfield, "Performance Analysis of Cellular Mobile Communication Systems with Dynamic Channel Assignment," *IEEE J. Select. Areas Commun.*, Vol. 7, No. 8, Aug. 1986, pp. 1172–1180.

[54] Cimini, L. J., and G. J. Foschini, "Distributed Algorithms for Dynamic Channel Allocation in Microcellular Systems," in *Proc. IEEE Vehicular Technol. Conf. VTC'92*, pp. 641–644, 1992.

[55] Sivarajanet, Kumar N., et al., "Dynamic Channel Assignment in Cellular Radio," in *Proc. IEEE Vehicular Technol. Conf. VTC'90*, pp. 631–635, Orlando, Florida, USA, May 1990.

[56] Beck, Reiner, and Herbert Panzer, "Strategies for Handover and Dynamic Channel Allocation in Micro-Cellular Mobile Radio Systems," in *Proc. IEEE Vehicular Tech. Conf. VTC'89*, pp. 178–185, 1989.

[57] Frodigh, Magnus, "Bounds on the Performance of DCA-Algorithms in Highway Microcellular Systems," *IEEE Trans. Vehicular Tech.*, Vol. 43, No. 3, Aug. 1994, pp. 420–427.

[58]   Madani, K., and A. H. Aghvami, "Performance of Distributed Control Channel Alloca-
       tion (DCCA) under Nonuniform Traffic Condition in Microcellular Radio Commun.,"
       in *Proc. IEEE Int. Conf. on Commun. ICC'94*, pp. 206–210, New Orleans, USA, May
       1994.

[59]   Leeet, Jongchan, et al., "Channel Allocation and Handover Schemes for Personal
       Communication Systems," in *Proc. IEEE Vehicular Tech. Conf. VTC'96*, pp. 943–947,
       Atlanta, GA, USA, 1996.

[60]   Cheng, Matthew M. L.,and J. C. I. Chuang, "Distributed Measurements-Based
       Dynamic Assignment for Personal Commun.," in *Proc. IEEE Vehicular Tech. Conf.
       VTC'95*, pp. 769–773, Chicago, Illinois, USA, June 25–28, 1995.

[61]   West, Kevin A., and Gordon L. Stuber, "An Aggressive Dynamic Channel Assignment
       Strategy for a Microcellular Environment," *IEEE Trans. Vehicular Tech.*, Vol. 43, No. 4,
       Nov. 1994, pp. 1027–1038.

[62]   Guerin, R. A., Ph.D. Thesis, Dept. of Electrical Engineering, CA Institute of Technol-
       ogy, Pasadena, CA., USA, 1986.

[63]   Hong, D., and S. S. Rappaport, "Traffic Model and Performance Analysis for Cellular
       Mobile Radio Telephone Systems with Prioritized and Nonprioritized Handoff Proce-
       dures," *IEEE Trans. Vehicular Technol.*, Vol. 35, No. 3, Aug. 1986, pp. 77–92.

[64]   Lin, Yi-Bing, Anthony R. Noerpel, and Daniel J. Harasty, "The Sub-Rating Channel
       Assignment Strategy for PCS Hand-Offs," *IEEE Trans. Vehicular Tech.*, Vol. 45, No. 1,
       Feb. 1996, pp. 122–129.

[65]   Noerpel, Anthony, and Yi-Bing Lin, "Handover Management for PCS Network," *IEEE
       Personal Commun. Mag.*, Vol. 4, No. 6, Dec. 1997, pp. 18–24.

[66]   Bing Lin, Yi, et al., "Queuing Priority Channel Assignment Strategies for PCS Hand-Off
       and Initial Access," *IEEE Trans. Vehicular Tech.*, Vol. 43, No. 3, Aug. 1994,
       pp. 704–712.

[67]   Tekinay, Sirin, and Bijan Jabbari, "A Measurement-Based Prioritization Scheme for
       Handovers in Mobile Cellular Networks," *IEEE Trans. Vehicular Tech.*, Vol. 10, No. 8,
       Oct. 1992, pp. 1343–1350.

[68]   Murase, Atsushi, et al. "Handover Criterion for Macro and Microcellular Systems," in
       Proc. IEEE Vehicular Technol. Conf.'91, pp. 524–528, St. Louis Missouri, USA, May
       19–22, 1991.

[69]   Lin, Yi-Bing, et al., "PCS Channel Assignment Strategies for Hand-Off and Initial
       Access," *IEEE Personal Commun.*, Vol. 1, No. 3, Third Quarter 1994, pp. 47–56.

[70]   Halpern, S. W., "Reuse Partitioning in Cellular Systems," in *Proc. IEEE Vehicular Tech-
       nol. Conf. VTC'83*, pp. 322–327, 1983.

[71]   Whitehead, F., "Cellular Spectrum Efficiency via Reuse Partitioning," in *Proc. IEEE
       Vehicular Technol. Conf. VTC'85*, 1985.

[72] Zander, J., and J. Frodigh, "Capacity Allocation and Channel Assignment in Cellular Radio Systems Using Reuse Partitioning," *Electronic Letters*, Vol. 28, No. 5, Feb. 1992, pp. 438–440.

[73] Madani, K., and A. H. Aghvami, "DCCA: A Distributed Control Channel Allocation Scheme for Microcellular Communication Networks," in *Proc. 7th IEEE Euro. Conf. On Mobile Personal Commun.*, 1993.

[74] Onoe, S., and S. Yasuda, "Flexible Re-Use for Dynamic Channel Assignment in Mobile Radio Systems," in *Proc. IEEE Int. Conf. on Commun. ICC'89*, pp. 472–476, Boston, MA, USA, June 11–14, 1989.

[75] Nettleton, R. W., "Traffic Statistics in a Self-Organizing Cellular Telephone System," in *Proc. IEEE Vehicular Technol. Conf. VTC'90*, pp. 305–310, Orlando, Florida, USA, May 6–9, 1990.

[76] Sengoku, M., et al., "Channel Assignment in Cellular Mobile Communication Systems and an Application of Neural Networks," *Trans. IECE*, Vol. J74-B-I, 1991, pp. 190–200.

[77] Takenaka, T., T. Nakamura, and Y. Tajima, "All Channel Concentric Allocation in Cellular Systems," in *Proc. IEEE Int. Conf. on Commun. ICC'93*, pp. 920–924, Geneva, Switzerland, May 23–26, 1993.

[78] Furukawa, H., and Y. Akaiwa, "Self Organized Reuse Partitioning, a Dynamic Channel Assignment Method in Cellular Systems," in *Proc. IEEE Vehicular Technol. Conf. VTC'93*, pp. 524–527, Secaucus, New Jersey, USA, May 18–20, 1993.

[79] Zander, Jens, "Generalized Reuse Partitioning in Cellular Mobile Radio," in *Proc. IEEE Vehicular Technol. Conf. VTC'93*, pp. 181–184, Secaucus, New Jersey, USA, May 18–20, 1993.

[80] Mihailovic, A., L. Ortigoza-Guerrero, and A. H. Aghvami, "Compact Pattern Based Dynamic Channel Assignment for Reuse Partitioning in Cellular Mobile Systems," in *Proc. Int. Conf. on Telecommun. ICT-98*, pp. 300–304, Porto Carras, Greece, June 21–25, 1998.

[81] Everitt, D., "Traffic Capacity of Cellular Mobile Commun. Systems," *Comp. Networks ISDN Systems*, Vol. 20, 1990, pp. 447–454.

[82] Frodigh, Magnus, "Reuse-Partitioning Combined with Traffic Adaptive Channel Assignment for Highway Microcellular Radio Systems," in *Proc. IEEE Vehicular Technol. Conf. VTC'92*, pp. 1414–1418, 1992.

[83] Salvalaggio, Tony, "On the Application of Reuse Partitioning," in *Proc. IEEE Vehicular Technol. Conf. VTC'88*, pp. 182–185, 1988.

[84] Kanai, T., "Autonomous Reuse Partitioning in Cellular Systems," in *Proc. IEEE Vehicular Technol. Conf. VTC'92*, pp. 782–785, 1992.

[85] Yum, Tak Shing Peter, and Kwan Lawrence Yeung, "Blocking and Handoff Performance Analysis of Directed Retry in Cellular Mobile Systems," *IEEE Trans. Vehicular Tech.*, Vol. 44, No. 3, Aug. 1995, pp. 645–650.

[86] Karlson, Johan, and Berth Eklundh, "A Cellular Mobile Telephone System with Load Sharing–An Enhancement of Directed Retry," *IEEE Trans. on Commun.*, Vol. 37, No. 5, May 1989, pp. 530–534.

[87] Madani, K., A. H. Aghvami, "Investigation of Handover in Distributed Control Channel Allocation (DCCA) for Microcellular Radio Systems," in *Proc. IEEE PIMRC'94*, pp. 160–163, 1994.

[88] Mihailescu, C., et al., "Dynamic Resource Allocation in Locally Centralised Cellular Systems," in *Proc. IEEE Vehicular Technol. Conf. VTC'98*, pp. 1695–1699, Ottawa, Canada, May 18–21, 1998.

[89] Chu, T. P., and S. S. Rappaport, "Overlapping Coverage and Channel Rearrangement in Microcellular Communication Systems," *IEEE Proc. Commun.*, Vol. 142, No. 5, Oct. 1995, pp. 323–332.

[90] Ganz, Aura, et al., "On Optimal Design of Multitier Wireless Cellular Systems," *IEEE Commun. Mag.*, Vol. 35, No. 2, Feb. 1997, pp. 88–93.

[91] Beraldi, R., et al., "Performance of a Reversible Hierarchical Cellular System," *Int. J. on Wireless Information Networks*, Vol. 4, No. 1, 1997, pp. 43–54.

[92] Lin, Y.-B., and A. Noerpel, "Modeling Hierarchical Microcell/Macrocell PCS Architecture," in *Proc. IEEE GLOBECOM'95*, pp. 405–409, Singapore, Nov. 13–17, 1995.

[93] Mihailescu, Claudiu, et al., "Analysis Of a Two-Layer Cellular Mobile Communication System," in *Proc. IEEE Vehicular Technol. Conf. VTC'97*, pp. 954–958, Phoenix, AZ, USA, May 4–7, 1997.

[94] Calin, Doru, and Djamal Zeghlache, "Traffic Modelling of a Hierarchical Cellular System in the Presence of Heterogeneous Calls," in *Proc. IEEE Vehicular Technol. Conf. VTC'97*, pp. 959–963, Phoenix, AZ, USA, May 4–7, 1997.

[95] Iera, A., et al. "Effective Radio Resource Assignment in Microcellular Systems with Cell Overlapping: The Reserve Channel Scheme," in *Proc. IEEE Vehicular Technol. Conf. '97*, pp. 964–968, Phoenix, AZ, USA, May 4–7, 1997.

[96] Fitzpatrick, Paul, et al., "Teletraffic Performance of Mobile Radio Networks with Hierarchical Cells and Overflow," *IEEE Trans. Vehicular Technol.*, Vol. 15, No. 8, Aug. 1995, pp. 1549–1557.

[97] Lo, Kuen-Rong, et al., "A Combined Channel Assignment Strategy in a Hierarchical Cellular Systems," in *Proc. IEEE Int. Conf. on Universal and Personal Commun., ICUPC'97*, pp. 651–655, San Diego, CA, USA, Oct. 1997.

[98] Wang, Li-Chun, and Gordon L. Stuber, "Architecture Design, Frequency Planning, and Performance Analysis for a Microcell/Macrocell Overlaying System," *IEEE Trans. Vehicular Tech.*, Vol. 46, No. 4, Nov. 1997, pp. 836–848.

[99] Brouetet, Jerome, et al., "Deployment of Multi-layer TDMA Cellular Network with Distributed Coverage for Traffic Capacity Enhancement," in *Proc. IEEE PIMRC'98*, Boston, MA, USA, 1998.

[100] Riva, G., "Preliminary Results on Traffic Overflow from Microcells to Macrocell," *IEEE Electronics Letters*, Vol. 28, No. 15, July 16, 1992, pp. 1462–1463.

[101] Frullone, M., G. Riva, P. Grazioso, and C. Carciofi, "Analysis of Optimum Resource Management Strategies in Layered Cellular Structures," in *Proc. Int. Conf. on Universal and Personal Commun., ICUPC'94*, pp. 371- 375, San Diego, CA, USA. 1994.

[102] Yeung, Kwan L., and Sanjiv Nanda, "Channel Management in Microcell/Macrocell Cellular Radio Systems," *IEEE Trans. Vehicular Technol.*, Vol. 45, No. 4, Nov. 1996, pp. 601–612.

[103] Chih-Lin, Larry J. Greenstein, and Richard D. Gitlin, "A Microcell/Macrocell Cellular Architecture for Low and High Mobility Wireless Users," *IEEE J. Select. Areas Commun.*, Vol. 11, No. 6, Aug. 1993, pp. 601–612.

[104] Lagrange, X., and P. Godlewski, "Performance of a Hierarchical Cellular Network with Mobility Dependent Hand-Over Strategies," in *Proc. IEEE Vehicular Technol. Conf. VTC'96*, pp. 1868–1872, Atlanta, GA, USA, April 28–May 1, 1996.

[105] Sung, Chi Wan, and Wing Shing Wong, "User Speed Estimation and Dynamic Channel Allocation in a Hierarchical Cellular System," in *Proc. IEEE Vehicular Technol. Conf. VTC'94*, pp. 91–95, Stockholm, Sweden, June 1994.

[106] Benveniste, Mathilde, "Cell Selection in Two-Tier Microcellular/Macrocellular Systems," in *Proc. IEEE GLOBECOM'95*, pp. 1532–1536, Singapore, Nov. 13–17, 1995.

[107] Shum, Kenneth W., and Chi Wan Sung, "Fuzzy Layer Selection Method in Hierarchical Cellular Systems," in *Proc. IEEE GLOBECOM'96*, pp. 1049–1053, London, UK, Nov. 18–22, 1996.

[108] Hu, Lon-Rong, and Stephen Rappaport, "Personal Communication Systems Using Multiple Hierarchical Cellular Overlays," *IEEE J. Select. Areas Commun.*, Vol. 13, No. 2, Feb. 1995, pp. 406–415.

[109] Rappaport, S. S., and L. R. Hu, "Microcellular Communication Systems with Hierarchical Macrocell Overlays: Traffic Performance Models and Analysis," *Proc. IEEE*, Vol. 82, No. 9, Sept. 1994, pp. 1383–1397.

[110] Jolley, W. M., and R. E. Warfield, "Modelling and Analysis of Layered Cellular Mobile Networks," *Teletraffic and Datatraffic in a Period of Change*, Vol. ITC-13, 1991, pp. 161–166.

[111] Unitel, "Idle Mode Cell Reselection for Microcells, in *ETSI GSM2*, Ronneby, Sweden, Sept. 1991.

[112] Karlsson, Robert, and Jens Zander, "Hierarchical Cell Structures for FRAMES Wideband Wireless Access," in *Proc. ACTS*, pp. 785–791, Spain, 1996.

[113] Scheibenbogen, Markus, et al., "Dynamical Channel Allocation in Hierarchical Cellular Systems," in *Proc. IEEE Vehicular Technol. Conf. VTC'96*, pp. 721–725, Atlanta, GA, USA, April 28–May 1, 1996.

[114] Lyberopoulos, G. L., et al., "The impact of Evolutionary Cell Architectures on Handover in Future Mobile Telecommunication Systesms," in *Proc. IEEE Vehicular Technol. Conf. VTC'94*, pp. 120–124, Stockholm, Sweden, June 1994.

[115] Ogawa, Keisuke, et al., "Optimum Multi-Layered Cell Architecture for Personal Communication Systems with High Degree of Mobility," in *Proc. IEEE Vehicular Technol. Conf. VTC'94*, pp. 644–648, Stockholm, Sweden, June 1994.

[116] Jabbari, Bijan, and Woldemar F. Fuhrmann, "Teletraffic Modelling and Analysis of Flexible Hierarchical Cellular Networks with Speed Sensitive Hand-off Strategy," *IEEE Trans. Vehicular Technol.*, Vol. VT-15, No. 8, Oct. 1997, pp. 1539–1548.

[117] Almgren, Magnus, et al., "Channel Allocation and Power Settings in a Cellular System with Macro and Micro Cells Using the Same Frequency Spectrum," in *Proc. IEEE Vehicular Technol. Conf.'96*, pp. 1150–1154, 1996.

3

DCA with Prioritization for Handover Calls in Microcellular Environments

3.1 Introduction

This chapter will show how combinations of CASs for initial access and handover in microcellular systems can decrease both the numbers of new calls blocked and the number of handover calls forced to terminate. To do this a teletraffic analysis is carried out for a microcellular system first using FCA and then using DCA.

The FCA system treats handover calls using both NPS and RCS. The analytical method used is valid for uniform and nonuniform traffic distributions. Then an extension (generalization) to the nonuniform compact pattern allocation algorithm [1] is presented as an application of this analysis.

Finally, based on this analytical method and on the extended nonuniform allocation concept, a modified version of the DCA strategy CPMCB is explained and combined with the NPS and the RCS strategies for handover for both uniform and nonuniform traffic.

The effect of mobility of users in the performance of the combination of the strategies is also assessed.

3.2 The Problem of Handovers in Microcellular Environments

A *handover*, of course, occurs when a call moves from one cell to another. The mobile terminal has to change its radio transmission from one BS to another. Clearly the more cell boundary crossings there are, the more forced termination of calls there will be. This will affect the design and performance of a cellular system. This is first discussed before analyzing performance in later sections of this chapter.

Recently, the demand for wireless communications has grown tremendously. Cell sizes have had to decrease to meet this demand, leading to the undesirable consequence of an increase in the number of handovers and in the probability that a call is forced to terminate. These microcells increase the capacity of systems by reusing the resources more intensively and are particularly useful in high-traffic demand areas.

Forced termination of ongoing calls is a less desirable event in the performance evaluation of a cellular communications network than the blocking of new calls because of the irritation it causes customers, so when a call is in progress in a cell great efforts are made to provide continuity to the current call when the user moves from one cell coverage to another.

Handover is an important function of mobility management. It is unique to cellular systems and especially crucial to support global roaming in personal communications services (PCS).

3.2.1 Types of Handover Strategies

So far three handover strategies have been proposed for PCS networks, classified depending on who initiates and executes the handover. The three strategies are:

1. Mobile-controlled handover (MCHO);
2. Network controlled handover (NCHO);
3. Mobile assisted handover (MAHO) [2, 3].

As stated in Chapter 2, several CAS strategies that deal with the initial access and handover problems have been reported in the literature. These CASs are:

1. Nonprioritized;
2. Reserved channel;
3. FIFO priority;
4. Measurement-based priority;
5. Subrating schemes.

The selection of a particular scheme is a trade-off between users' quality of service requirements and network operating costs. Each of these strategies can be implemented with FCA or any other type of CAS (i.e., DCA). However, handover strategies do have some trade-offs when implemented and, while they can decrease the number of calls that are forced to terminate and so improve the call incompletion rate, at the same time, they increase the blocking probability of new calls (if they are not used with a DCA). Handover strategies are also affected by user mobility.

3.2.2 Methods of Decreasing New Call Blocking

Until recently research on CASs had focused mainly on decreasing the blocking probability of new calls [4–9], but now new algorithms have been proposed to cope with the handover problem as well [10–16]. Among channel assignment strategies, channel borrowing and DCA have shown good performance in decreasing the blocking probability of new calls when the offered load to the sys   both under uniform and nonuniform traffic patterns. A combination of   wo, a DCA strategy using maximized channel borrowing, was presented in [17]. This strategy takes advantage of the compact pattern concept [1] to decrease the blocking probability of new calls by always assigning a channel to a pattern that improves the system performance. Channel borrowing and channel reallocation strategies are also used to minimize the blocking probability for each cell of the system. CPMCB has outperformed FCA, BCO, and BDCL in two different evaluation environments [17, 18]. Its feasibility when applied to TDMA systems has also been studied [19]. Following the concepts stated in [13], CPMCB could be classified as an *interference-free, timid, not conditioned class* of DCA strategy and also, according to [13], as a "complex" strategy.

3.2.3 Analysis Later in This Chapter

In this chapter, an extension to the nonuniform compact pattern allocation algorithm [1] is explained first, as an example of how to carry out such problems. This modification takes into account not only the new call arrival rate but also the handover arrival rate in each cell. Only the new call arrival rate can realistically be supposed to be known; a method of deriving, from it, the handover arrival rate for any kind of traffic is presented. Then, this extended concept is applied to develop a modified version of CPMCB [17] to minimize both the blocking rate of new calls (p_b) and the handover failure rate (p_h). Also, NPS and RCS are combined with CPMCB. Then, at the end of the chapter, simulation results are given showing the performance of this modified version of CPMCB in terms of the forced termination probabilities and the probability that a call is not completed. User mobility is also taken into account in the results.

It is also a concern of this chapter to show that, to some extent, CPMCB with a modified cost function can act as a self-prioritized handover strategy and to compare CPMCB with a well-established reference (in this case the FCA strategy), since in previous research CPMCB was shown to perform better than some channel borrowing and DCA strategies [17–19].

3.3 Analytical Method for Calculating System Probabilities p_b, p_h, p_{ft} and p_{nc}

In this section analytical expressions are derived for these probabilities for a cellular system using FCA with NPS and RCS priority schemes, knowing just the new call arrival rates. Simplified expressions are given also for a uniform system with all cells that have the same statistical properties. Finally a procedure for obtaining numerical results from these expressions is given.

3.3.1 Assumptions and Symbols

The probabilities being derived are:

- *Blocking probability, p_b* : the probability that a new user finds all channels busy in a certain cell;
- *Hand-off failure probability, p_h*: the probability that a handover call finds all channels busy on its arrival at its target cell;
- *Probability of forced termination, p_{ft}* : the probability that a call originally accepted by the system is interrupted during its process due to a handover failure;[1]
- *Probability that a call is not completed, p_{nc}* : the probability that a call is not completed either by blocking (for a new call attempt) or by forced termination (for a handover call).

To derive methods for calculating the sets of probabilities of interest for this uniform system, a single cell is considered and the following quantities are used.

---

1. Note that in real systems a handover failure does not automatically mean that the call will drop. In the majority of cases the call will remain (degraded) on the original cell and the network will continue to attempt handover until the signal strength or quality falls below predefined thresholds. However, for analysis purposes, we assume that immediately after a user crosses the cell boundary it will perform a handover and, if there is no channel available in the target cell to attend the call, it will be dropped (forced to terminate).

- *New call arrival rate, λ_0*: it is assumed that new call attempts in the cell follow a Poisson process and that λ_0 represents the new call arrival rate in calls/s.

- *Handover call arrival rate, λ_{hi}*: It is assumed that handover arrivals, from the six neighboring cells, follow a Poisson process and that λ_{hi} represents the handover call arrival rate into a given cell (when the traffic in the system is uniform this will be the same as the hand-off departure rate λ_{ho} from the cell).

- *Dwell time, t_m*: If the mobile is given a channel by the cell, the mobile remains in the cell's coverage area for a period of time, called the dwell time, before it moves out of the cell. t_m is a random variable that could have a general probability distribution function F_m with mean $1/\eta$ as proposed in [20], but, as is usual, t_m is assumed to be exponentially distributed here with probability density function $f_m(t_m) = \eta e^{-\eta t}$ and mean $1/\eta$.

- *Call holding time, t_c* : It is also supposed that t_c is exponentially distributed with density function $f_c(t_c) = \mu e^{-\mu t}$ with mean $1/\mu$. The call holding time is the amount of time that the call would remain in progress if it continued to completion without forced termination due to handover failure [21].

- *Channel occupancy time, t_{co}*: If a mobile is given a channel, this channel would be released either by the completion of the call in the cell or by a handover process to a neighboring cell. In this way, the channel occupancy time is the smaller of the call holding and dwell times. This is a result of the lack of memory of the negative exponential distribution. The probability density function of the channel occupancy time distribution is given by [20]

$$f_{co}(t) = \int_{t_c=t}^{\infty} f_c(t_c)f_m(t)\,dt_c + \int_{t_m=t}^{\infty} f_c(t)f_m(t_m)\,dt_m = (\mu + \eta)e^{-(\mu+\eta)t} \qquad (3.1)$$

3.3.2 p_b and p_h with NPS

In this section, the blocking probability is derived for FCA with NPS.

NPS can be modeled by a Markov process with $s + 1$ states, where s is the number of available channels in a certain cell. In this case, new calls and handover arrival calls can use all s channels as long as they are idle. So that for $0 \leq j < s$, the next arrival (new call or handover arrival call) can occupy the next idle channel. Let P_j be the statistical-equilibrium probability of the new call arrival encountering j busy servers. Then for $0 \leq j < s$, the transition rate from state P_j to P_{j+1} is given by $\lambda_0 + \lambda_{hi}$ and a transition from state P_{j+1} to P_j is given with the rate $(j + 1)(\mu + \eta)$ because the channel occupancy time is exponentially distributed. Suppose that the s channels are all busy. Then if a new call arrives at that moment, the call will be blocked. Also if a handover call arrives, it is forced to terminate and a handover failure is produced. From the state diagram of Figure 3.1, it is easy to find the steady-state probability P_j

$$P_j = \frac{\left(\dfrac{\lambda_o + \lambda_{hi}}{\mu + \eta} \right)^j}{j!} P_0 \tag{3.2}$$

Then using the normalization equation

$$\sum_{j=0}^{s} Pj = 1 \tag{3.3}$$

gives

$$P_j = \frac{\dfrac{\left(\dfrac{\lambda_o + \lambda_{hi}}{\mu + \eta} \right)^j}{j!}}{\displaystyle\sum_{k=0}^{s} \dfrac{\left(\dfrac{\lambda_o + \lambda_{hi}}{\mu + \eta} \right)^k}{k!}} \tag{3.4}$$

where k is merely a variable of summation. Then p_b can be found when $j = s$ in (3.4), as

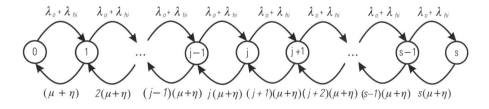

Figure 3.1 The state diagram for the NPS. ©1999 IEEE. Reprinted, with permission, from [22].

$$p_b = P_s = \dfrac{\dfrac{\left(\dfrac{\lambda_o + \lambda_{hi}}{\mu + \eta}\right)^s}{s!}}{\displaystyle\sum_{k=0}^{s} \dfrac{\left(\dfrac{\lambda_o + \lambda_{hi}}{\mu + \eta}\right)^k}{k!}} \qquad (3.5)$$

and, since there is no priority, $p_h = p_b$.

3.3.3 p_b and p_h with RCS

In this scheme, some of the s channels are reserved for handover arrival calls only.

Suppose that Ch channels are reserved out of the s total available channels. Then the state diagram that describes the system performance is shown in Figure 3.2. If there is a new call arrival, it will be attended if the number of busy channels at that time are less than n, where $n = s - Ch$; otherwise the call will be blocked. Handover arrivals will only be blocked if the number of busy channels in the target cell is equal to s, therefore producing a forced termination.

This case can also be modeled by a Markov process with $s + 1$ states, but a distinction must be made between the system behavior when the number of busy channels is less than n and when it is more than n but less than s. Let P_j be the statistical-equilibrium probability of j busy servers. Then for $0 \le j < n$, a transition rate from state P_j to P_{j+1} is given by $\lambda_0 + \lambda_{hi}$ and a transition from state P_{j+1} to P_j is given with rate $(j + 1)(\mu + \eta)$ because the channel occupancy time is exponentially distributed. For $n \le j < s$, a transition rate from state P_j to P_{j+1} is given by λ_{hi} and a transition from state P_{j+1} to P_j is given with the rate $(j + 1) \times (\mu + \eta)$ because new call arrivals are not allowed to occupy a channel when there are n or more busy channels in the cell. Let us say that the n channels are busy. If a new call arrives at that moment, the call will be blocked, but if a handover arrival occurs, it will be attended. Considering the state diagram of Figure 3.2, the steady-state probability, P_j is

$$P_j = \dfrac{\left(\dfrac{\lambda_o + \lambda_{hi}}{\mu + \eta}\right)^j}{j!} P_0 \qquad\qquad 0 < j \le n \qquad (3.6)$$

and

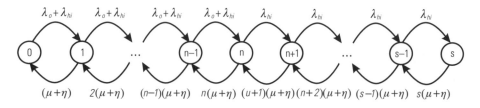

Figure 3.2 The state diagram for the RCS. ©1999 IEEE. Reprinted, with permission, from [22].

$$P_j = \frac{\left(\lambda_{hi}\right)^{j-n}\left(\lambda_o + \lambda_{hi}\right)^n}{\left(\mu + \eta\right)^j j!} P_0 \qquad n < j \le s \tag{3.7}$$

taking (3.3) into account gives

$$P_0 = \left[\sum_{j=0}^{n} \frac{\left(\dfrac{\lambda_o + \lambda_{hi}}{\mu + \eta}\right)^j}{j!} + \sum_{j=n+1}^{s} \frac{\left(\lambda_{hi}\right)^{j-n}\left(\lambda_o + \lambda_{hi}\right)^n}{j!\left(\mu + \eta\right)^j} \right]^{-1} \tag{3.8}$$

In this case, the probabilities p_b and p_h are given by

$$p_b = \sum_{j=n}^{s} P_j \tag{3.9}$$

and

$$p_h = P_s \tag{3.10}$$

3.3.4 Probabilities of Forced Termination and Call Not Completed, p_{ft} and p_{nc}

Here a general expression is derived to calculate the forced termination probability p_{ft} and the probability that a call is not completed p_{nc} in an arbitrary regular network of hexagonal cells with an arbitrary traffic pattern (i.e., uniform or nonuniform).

The blocking probability in cell k is given by $p_b(k)$ and the probability $S(k)$ that a new call origination is accommodated by the system in cell k is [21]

$$S(k) = \frac{(1 - p_b(k))}{(1 - p_{bT})} \tag{3.11}$$

where p_{bT} is the average new call blocking probability over all cells, given by

$$p_{bT} = \frac{\sum_{k=1}^{N} p_b(k)}{N} \tag{3.12}$$

where N is the total number of cells in the system. If a call is being serviced by cell k, then the probability $a(k)$ that this call makes a handover to one of its neighboring cells, say cell i, and is successfully accommodated is

$$a(k) = \frac{\eta}{\mu + \eta} \sum_{i=1}^{6} q(k, i)(1 - p_h(i)) \tag{3.13}$$

where $q(k, i)$ is the probability that a user with an engaged call in cell k leaves the cell (handover departure) by the ith side and $p_h(i)$ is the handover failure probability in the target cell i. The multiplier takes into account the relative values of dwell time and call holding time (see start of Section 3.2 for details).

The probability that a call currently served by cell k will make a handover attempt to cell i and be denied access to a channel in the target cell i is given by

$$b(k) = \frac{\eta}{\mu + \eta} \sum_{i=1}^{6} q(k, i) p_h(i) \tag{3.14}$$

With independent handover attempts, the probability that a mobile currently served by a cell in the system is forced to terminate after 0, 1, … , n successful handovers can be approximated by [21]

$$E(k) = b(k) + a(k)b(k) + a^2(k)b(k) + \cdots + a^n(k)b(k) + \cdots = \frac{b(k)}{1 - a(k)} \tag{3.15}$$

Substituting (3.13) and (3.14) into (3.15) yields

$$E(k) = \frac{\eta \beta(k)}{\mu + \eta - \eta \alpha(k)} \tag{3.16}$$

where

$$\alpha(k) = \sum_{i=1}^{6} q(k,i)(1 - p_b(i)) \tag{3.17}$$

and

$$\beta(k) = \sum_{i=1}^{6} q(k,i) p_h(i) \tag{3.18}$$

So, finally, the p_{ft} is

$$p_{ft}(k) = S(k)E(k) = \frac{(1 - p_b(k))}{(1 - p_{bT})} \frac{\eta\beta(k)}{\mu + \eta - \eta\alpha(k)} \tag{3.19}$$

where $S(k)$ is the probability that the call was accepted initially (i.e., not blocked).

In [20] it is shown that, when $F_m(t)$ is exponentially distributed,

$$p_{nc}(k) = 1 - \frac{1 - p_b(k)}{1 + \dfrac{\eta p_h(k)}{\mu}} \tag{3.20}$$

To calculate these two probabilities, values for p_b and p_h must be known. p_b can be measured, but evaluating p_h requires calculating the handover departure rate of each cell k, $\lambda_{ho}(k)$, knowing only the new call arrival rate to the same cell k, $\lambda_o(k)$. To calculate $\lambda_{ho}(k)$, it is considered that the probability that a user in cell k makes a handover is given by

$$c = c(k) = \frac{\eta}{\eta + \mu} \tag{3.21}$$

while the probability of doing a successful handover is given by (3.13). Once again, with independent handovers, the probability that a call makes a handover after it has done 0, 1, ... , n previously successful handovers is

$$D(k) = c + a(k)c + a(k)^2 c + \cdots + a(k)^n c + \cdots = \frac{c}{1 - a(k)} \tag{3.22}$$

so that

$$D(k) = \frac{\eta}{\mu + \eta - \eta a(k)} \qquad (3.23)$$

The handover departure rate from cell k, $\lambda_{ho}(k)$, can then be found by

$$\lambda_{ho}(k) = (1 - p_b(k))\, D(k)\lambda_o(k) \qquad (3.24)$$

Substituting (3.23) and (3.17) into (3.24) gives

$$\lambda_{ho}(k) = \frac{(1 - p_b(k))\lambda_o(k)\eta}{\mu + \eta - \eta \displaystyle\sum_{i=1}^{6} q(k,i)(1 - p_b(i))} \qquad (3.25)$$

Once the handover departure rate from cell k is known, the handover arrival rate can be found in each cell as

$$\lambda_{hi}(k) = q(k,i)\sum_{i=1}^{6} \lambda_{ho}(i) \qquad (3.26)$$

Next assume the cells are all uniform so that users leave the cell by each of its six sides with equal probability, $q(k, i) = 1/6$. Then

$$\lambda_{hi}(k) = \frac{1}{6}\sum_{i=1}^{6} \lambda_{ho}(i) \qquad (3.27)$$

But if the traffic in all cells is uniform, the variables can be simplified because handover is equally likely to each of its six neighbors. This gives

$$\lambda_{ho}(k) = \lambda_{hi}(k)$$
$$p_b(k)p_{bT} = p_b$$

so

$$S = S(k) = 1$$
$$p_b(i) = p_b$$

Hence, (3.13) and (3.14) become

$$a = \frac{\eta}{\mu + \eta}(1 - p_h) \tag{3.28}$$

and

$$b = \frac{\eta}{\mu + \eta}p_h \tag{3.29}$$

Then

$$E = \frac{\eta p_h}{\mu + \eta p_h} \tag{3.30}$$

and

$$P_{ft} = \frac{1}{\dfrac{\mu}{\eta p_h} + 1} \tag{3.31}$$

Note that E and p_{ft} are the same as you would expect for a system with uniform handover probability.

Finally note that if $\mu \gg \eta$, calls almost always end before the mobile leaves the cell and therefore p_{ft} is very small, decreasing as μ increases. On the other hand, if $\mu \gg \eta$, almost all users will leave the cell before completing the call and the number of forced termination calls will increase as η increases.

3.3.5   Calculation of Probability Values

To find the output measures p_b, p_h, p_{ft}, and p_{nc} in a network with an arbitrary traffic pattern, an iteration method is used in much the same manner as described and proposed in [23]. An important difference here is that it is dealing with nonuniform traffic and there is more than one equation to solve. As in [23], the procedure converged for all the cases studied in this Chapter. The procedure is as follows.

      Input: s, n, μ, η, and $\lambda_o(k)$

      Output: p_b, p_h, p_{ft}, and p_{nc}

      Step 0: $\lambda_{ho}(k) = 0.2 * \lambda_o(k), \delta = 1$

Step 1:   If $|\delta| < 0.0001$ for all k go to Step 4

Step 2:   Compute $P_o(k)$ and $P_j(k)$ according to the handover strategy used.

Compute $p_b(k)$ and $p_b(k)$ according to the handover strategy used.

Step 3:   Compute the new value for $\lambda_{ho}(k)$ using (25)

Compute the new value for $\lambda_{h1}(k)$ using (27)

Let δ be the difference between the old $\lambda_{ho}(k)$ and the new $l_{ho}(k)$.

Go to Step 1.

Step 4:   Compute p_{ft} using (19) and p_{nc} using (20)

3.4 CAS

We now look at a modified DCA strategy that increases channel utilization. It is CPMCB and is described fully in [17], but a shorter description is provided here to enhance clarity. Then the modifications to the nonuniform compact pattern allocation algorithm and further modifications to CPMCB are described.

CPMCB is divided in two main phases: *channel allocation* and *channel releasing*. Channel allocation is further divided into nonuniform channel allocation and borrowed channel allocation.

3.4.1 Channel Allocation Description

This first phase of CPMCB, called channel allocation, is further divided into a nonuniform channel allocation and a borrowed channel allocation.

3.4.1.1 Nonuniform Channel Allocation

Consider an environment of W cells and assume that cell k is part of W ($k \in W$) and is operating with a compact pattern technique. A compact pattern is one that allocates any new channel in such a way that takes advantage of the reuse distance to allow as many identical channels to be packed in as possible; that is, the system is compacted. This allows more calls to be attended and so maximizes system capacity. Then when a channel (x) is needed for a new call there are always two well-compacted patterns (Y_1 and Y_2), clockwise and counterclockwise, to which the call could be assigned; the one that gives the largest reduction in the overall blocking rate of the system is chosen for use. Thus, if $\lambda_o(k)$ is the offered traffic in Erlangs to cell k and s_k is the current number of channels allocated to cell k, the blocking rate in cell k is obtained for Poisson call arrivals by [1, 8, 17]

$$R_k\left(s_k\right) = \Delta_o(k) \left[\sum_{p=0}^{s_k} \frac{\Delta_o(k)^p}{p!}\right]^{-1} \frac{\Delta_o(k)^{s_k}}{s_k!} \qquad (3.32)$$

The compact pattern Z that gives the larger reduction of the blocking rate is obtained by the decision rule [8,17]

$$Z = \begin{cases} Y_1 \; if \left[\sum_{k \in Y_1} R_k\left(s_k\right) - \sum_{k \in Y_1} R_k\left(s_k + 1\right)\right] > \left[\sum_{k \in Y_2} R_k(s_k) - \sum_{k \in Y_2} R_k\left(s_k + 1\right)\right] \\ Y_2 \; otherwise \end{cases} \qquad (3.33)$$

where Y_1 and Y_2 are the two compact patterns associated with each cell. The allocation of each channel to one of the compact patterns is made without changing the allocation of the other channels already given to the same compact pattern. This nonuniform channel allocation is applied whenever a channel is free and needs to be assigned to a compact pattern. In fact, this nonuniform allocation technique improves the system performance when nonuniform traffic is taken into account; indeed, if traffic is uniform, the channel assignment in the system is also uniform [1, 17].

3.4.1.2 Borrowed Channel Allocation

When a new call arrives an idle nominal channel is assigned to the call. If there is not a nominal idle channel available, a borrowed channel is tried (selected from a set of channels not in use in the interference neighborhood of the cell where the arrival occurred; the interference neighborhood for a given cell is the set of cells that do not fulfill the reuse distance constraints) and an optimal channel allocation algorithm is obtained by a minimum cost function. This cost function always tries to assign the least used channel in the system by asking for the current use of the channel in the cochannel cells of the lending cell.

In CPMCB the classical channel borrowing concept [4] is extended by defining five possible statuses in which a channel could be in a certain cell. To define channel status, one, two, or four cells are taken into account: the current cell (where the arrival occurs), one neighboring cell (the lending cell), and the two closer cochannel cells of the lending cell (from those surrounding the borrower cell). The possible channel statuses of channel x in a certain cell are:

- *Free* if the channel is idle;

- *Occupied* if the channel is being used by a local call;

- *Locked* when the channel is being locked by a borrowing in one of its cochannel cells;

- *Double-locked* when the channel is being lent by two cochannel cells or by only one of them and by the cell itself;

- *Triple-locked* when the channel is being lent by three cochannel cells or in two cochannel cells and in the cell itself;

- *Borrowed* when channel *x* is being borrowed by a neighboring cell.

Based on the definition of the possible statuses of a channel, in [17] it was found that there exist as many as 27 ways in which a channel can be borrowed while fulfilling the cochannel interference constraints. A detailed explanation of this phase and the relationship between the channel statuses with the cost functions can be found in [17].

3.4.2 Channel Releasing

Phase two of the algorithm, called channel releasing, is in charge of the reallocations after a call has finished.

Whenever a call in a nominal channel is over, a reallocation is made in order to reassign this channel to a call attended in a borrowed channel. Channel reallocation is carried out once a call is over, therefore, at most, one channel is reassigned at each repacking. As in the first phase of CPMCB, a cost function is used to determine which borrowed channel should be released [17]. The cost function always tries to release the least lent channel in order to increase the number of free channels in the system. If more than one channel meets the condition, a choice is made randomly. Phase two also has the task of changing the channels to a different compact pattern whenever a channel is released, no reallocation procedure has occurred, and no other cell in the system is using it.

3.4.3 Modifications to CPMCB

In this section modifications to CPMCB that can enhance its behavior when handover is considered are explained.

For our purposes, the only phase or subphase that has to be changed is the nonuniform channel allocation, which is based on the compact pattern concept (Section 3.4.1.1). What is described here is a modification to the cost function that selects the best compact pattern to which a channel should be assigned as a nominal channel. The cost function is given by (3.32) and (3.33).

In this new version of CPMCB, if a channel is going to be assigned from the central pool to cell k, a new decision rule is taken into account to decide to which compact pattern the channel will be placed (every cell has only two patterns: the clockwise and the counterclockwise). The compact pattern that gives the largest reduction in the overall new call blocking probability (p_b) and in the overall handover arrival blocking probability (p_h) (and therefore in p_{fi} and p_{nc}) is selected. Taking into account the new call arrival rate $\lambda_o(k)$, the handover arrival rate $\lambda_{hi}(k)$ and the number of currently allocated channels s_k in cell k the new cost function is given by

$$\Omega_k(s_k) = \lambda_o(k)p_b(k) + \lambda_{hi}(k)p_h(k) \qquad (3.34)$$

The compact pattern Z that gives the largest reduction in the overall p_b and p_h is obtained by the decision rule

$$Z = \begin{cases} Y_1 & if \left[\sum_{k \in Y_1} \Omega_k(s_k) - \sum_{k \in Y_1} \Omega_k(s_k + 1) \right] > \left[\sum_{k \in Y_2} \Omega_k(s_k) - \sum_{k \in Y_2} \Omega_k(s_k + 1) \right] \\ Y_2 & otherwise \end{cases} \qquad (3.35)$$

$\lambda_{hi}(k)$ is calculated according to the algorithm described in Section 3.3.5, so a correct estimate of and balance between new call arrivals and handover arrivals is made to assign a suitable number of channels to each cell in the system increasing the prioritization to handover calls and taking into account user mobility. This creates adaptability to changes in user mobility. With this new cost function, one can easily see that, when the traffic in the system is uniform, the channel allocation is uniform as well. In the earlier CPMCB version only the new call arrival rate was taken into account in deciding the pattern to which a channel should be assigned. With this enhancement to the nonuniform compact pattern allocation algorithm, the performance of other CASs such as FCA, BCO [5], LODA [4], and BDCL [4] as shown in [1] could be further improved. The same could be expected for compact pattern–based DCA (CP-based DCA) [8].

3.4.4 Practical Implementation of CPMCB

The number of calls per unit of time in every cell is the a priori information needed by the BSs and the relevant switching centers to execute the first phase of CPMCB. The allocation of channels to the compact patterns will depend only on the current offered load to the cells in the network. This load will have to be measured and updated periodically. However, this should not be a difficult task because the number of calls per unit of time can be found relatively easily in cellular networks and statistics reporting this value are generated continuously during the day in these networks.

It is normally assumed that every BS can instantaneously know the channel utilization in neighboring BSs so that channel synchronization (in channel allocation and channel releasing phases) would not be necessary. However, the practical implementation of DCA strategies requires a similar approach to that described in [13], which can always be used to account for finite propagation delays in the information exchange among BSs and to avoid colliding decisions in carrier acquisition.

In [13] it is proposed to divide the time axis into time intervals of equal duration, called decision time slots (DTS). These intervals are greater than, or equal to, the time necessary for a BS to update its channel utilization map (i.e., the channels that are being used in the interference neighborhood). Every DTS is associated with a compact pattern in a cyclic fashion (therefore, a cell will be assigned two DTS because every cell belongs to two compact patterns). Then, based on the channel allocation algorithm, the cells associated with a given compact pattern decide which channel to acquire at the end of the DTS only. Channel releases can be carried out regardless of the current DTS since they cannot generate any collisions.

There are several ways in which the BSs can get information about channel utilization; through a fixed signaling network connecting the BSs and/or the relevant switching centers or simply by listening to the carriers emitted by neighboring BSs [13].

CPMCB is a centralized algorithm that would need a central controller to make decisions about which compact pattern a channel should be assigned, for example. For these purposes, a dedicated fixed signaling network is the best possible option to communicate between BSs and the switching centers. The centralized feature of this strategy avoids colliding decisions in carrier acquisitions but increases the signaling load considerably.

3.5 Evaluation Environment

To allow meaningful simulations of the effect of the strategies, a large realistic microcellular network must be used. To achieve adequate realism, the network is based on the size, shape and measured traffic pattern of an actual network with suitable simplifications to allow simulation. A wraparound topology in the simulations consisting of 49 cells is proposed. This eliminates the boundary effect that occurs in an unwrapped topology and ensures statistically good results. The reuse pattern in the simulation is 7 for FCA and CPMCB, although it could have been any other value. An hexagonal cell geometry is used, and seven cochannel cells for each of the central cluster cells are kept for possible cochannel interference evaluation, since the first tier of cochannel interferers produce most of this type of interference. The mobility behavior of mobiles in the simulation is described by a two-dimensional random walk as used in [23]. In this model, a mobile stays in the coverage area of a cell for a period of time (dwell time) that has an exponential distribution with mean $1/\eta$. Then the mobile moves to one of the six possible neighbors with probability 1/6.

The proposed simulation environment is shown in Figure 3.3. There are seven clusters of seven cells each with the central cell in each cluster marked with the enlarged numbers 10, 20, ... , 70, respectively. The six other cells in each cluster are numbered counterclockwise following the number in the central cell. The white cells are the actual 49 cells used in the simulations, and the shaded cells are used to create the wraparound topology. Placing the shaded cells at the border of the 49 cell system ensures that each cell in the system has six neighboring cells and that no duplicate neighbor is created.

In Figure 3.4, a random nonuniform spatial traffic pattern taken from real data is shown. The number in each cell represents the offered load and ranges from 0.5 Erlangs to 36 Erlangs. This traffic pattern is used to evaluate

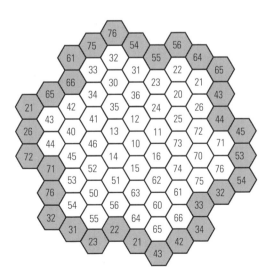

Figure 3.3 Evaluation environment used (with cell numbering). ©1999 IEEE. Reprinted, with permission, from [22].

the strategies considered in this chapter. The data was taken from the core area of a real cellular system formed by a HCS consisting of microcells overlaid by a set of macrocells and applied to a simplified system with a regular structure formed of a grid of hexagonal cells (like that shown in Figure 3.3) to allow analysis and simulation to be carried out straightforwardly. The traffic pattern was obtained from the busy hour of the busy day of the week. The 49 active cells in our system include all the actual microcells in the real system, which were less than 49 in number and all hexagonal (i.e., surrounded by six adjacent cells). The real cells were matched to our cells in position and, consequently, in load. For these cells the traffic load used was that of the corresponding actual cell but simplified because the actual cells were sectored so that the sum of the sector loads was used as the total cell load and assumed to be distributed uniformly throughout the cell. In the few positions in our pattern where no real microcell existed to provide data, a loading figure was obtained from the overlying macrocell by taking a fraction of its load proportional to relative areas. None of the other microcells had any load added from its macrocell. As a result, a realistic loading pattern was obtained (as in Figure 3.4) in a system that is analytically and computationally manageable.

This approach to model a nonuniform load is, however, quite static and cannot describe the transients in time (and variations in space) that may happen in a real cellular network. There, at certain hours during the day, some cells may experience more traffic load than in the peak hour defined on the overall

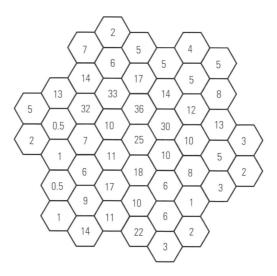

Figure 3.4 Nonuniform traffic pattern used (load in Erlangs). ©1999 IEEE. Reprinted, with permission, from [22].

network and the dynamics of the changes even during a limited period of time may significantly alter the performance of the network.

3.6 Examples and Discussions

In this section the results obtained for the network proposed in Figure 3.3 are shown when uniform and real nonuniform traffic patterns are taken into account. Results are shown, both analytical and simulated, for the FCA strategy with NPS for handovers (FCA-NPS) and with RCS for handovers with $Ch = 1$, and $Ch = 2$ reserved channels (FCA-RCS1 and FCA-RCS2, respectively). All simulation results are plotted with a 97% confidence interval. For a uniform traffic simulation all the sets of curves presented in Figures 3.5 to 3.12 agree completely with analytical results, whereas for those obtained with nonuniform traffic, small disagreements exist. The CPMCB strategy is also evaluated with NPS (CPMCB-NPS) and RCS schemes with $Ch = 1$ and $Ch = 2$ reserved channels (CPMCB-RCS1 and CPMCB-RCS2). In CPMCB the number of reserved channels, as in the FCA strategy, is the number of channels that are reserved exclusively for handover call arrivals and is constant. All the sets of probabilities presented were obtained by simulations averaged on all the 49 cells due to the wraparound topology used. To reflect the PCS environment, the total number of channels in the system was chosen to be 70; therefore, the number of channels per cell is $s = 10$ when using FCA. For all cases the mean call holding time is $1/\mu = 3$ min. The mobile mean dwell time $1/\eta$ is initially taken as 3 min and then varied by varying η from $\eta = 0.5\mu$ to $\eta = 6\mu$ to evaluate the influence of increasing the mobile speed (decreasing the cell dwell time). Note that channels and not carriers are considered in this chapter due to the flexibility that this gives for future work. In general terms, a channel could be a frequency in a FDMA system, a time slot within a frame in a TDMA system, a code in a WB-CDMA-TDD system, a fraction of a time slot or a code associated with a time slot in a hybrid TD/CDMA system, or a combination thereof (some constraints may occur in some of the different cases depending on the technology used and the implementation).

3.6.1 Uniform Traffic

In this section, the effects of constant and changing mobility of users is taken into account by system performance evaluation when the traffic is uniformly distributed.

Both FCA and CPMCB CASs are considered, each with various prioritization strategies, NPS and RCS. Performance results are given, showing how the four system probabilities (p_b, p_h, p_{ft}, and p_{nc}) vary with load and dwell time (mobility). They were obtained by both simulation and analysis (as explained in Section 3.3.5), although these produced such similar values that they are not presented separately in the graphs in Figures 3.5 to 3.12.

A similar evaluation is presented for nonuniform traffic in Section 3.6.2 but with sufficient differences in the simulated and analytical results to justify displaying both.

3.6.1.1 Uniform Traffic with Fixed, Constant Mobility; Varying Uniform Load

Performance values are shown as functions of offered load and user speed in Figures 3.5 to 3.8 for uniform traffic distribution and constant mobility.

The system parameters for call holding times used in the evaluations were both set at $1/\mu = 1/\eta = 3$ min with offered load varying from 2 to 12 Erl/cell. For all these strategies it can be seen that, with little exception, CPMCB-based strategies perform much better than those using FCA. In particular:

1. Figure 3.5 shows that the blocking probability (p_b) for new call attempts at the same traffic load with any channel assignment strategy is smaller using NPS than with RCS. This observation is consistent with previous studies that determined giving priority to handover calls increases the blocking probability. The blocking probability of new calls for CPMCB is always smaller than for FCA over all the evaluation range. An increase in the system capacity by a factor of 1.55 is observed for CPMCB-NPS in comparison with FCA-NPS at a blocking rate of 0.01.

2. Figure 3.6 shows the handover failure probability in a cell. It can be seen that giving priority to handover calls has a greater impact in FCA than in CPMCB because CPMCB is a strategy that has already included an inner prioritized cost function that balances p_b and p_h. That is why no significant difference is observed between the curves for CPMCB with zero, one, and two reserved channels. However, CPMCB has the same performance as FCA with two reserved channels for handover arrival calls at a blocking rate of 0.01.

3. In Figure 3.7, p_{ft} follows the same performance as p_h, since these two probabilities are strongly linked.

4. The results presented in Figure 3.8 confirm that CPMCB-NPS has a better performance than any combination FCA-handover scheme in terms of p_{nc}. The probability that a call is not completed (p_{nc}) gives an overall characterization of the user perceived grade of service. CPMCB provides a lower p_{nc} than FCA for small and moderate

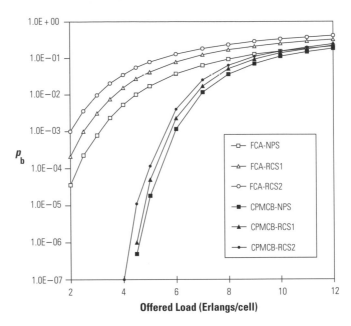

Figure 3.5 Blocking probability p_b with $1/\eta = 1/\mu = 3$ min and uniform traffic. ©1999 IEEE. Reprinted, with permission, from [22].

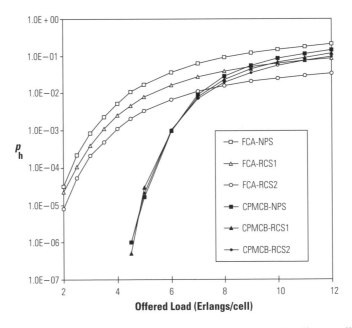

Figure 3.6 Handover failure probability p_h with $1/\eta = 1/\mu = 3$ min and uniform traffic. ©1999 IEEE. Reprinted, with permission, from [22].

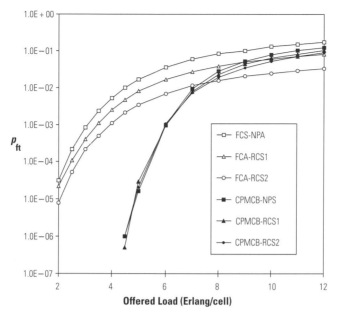

Figure 3.7 Forced termination probability p_{ft} with $1/\eta = 1/\mu = 3$ min and uniform traffic. ©1999 IEEE. Reprinted, with permission, from [22].

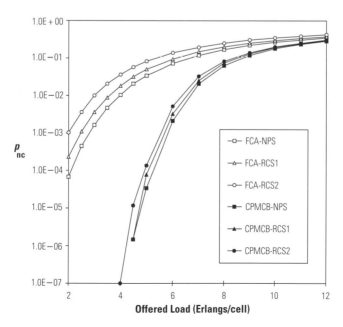

Figure 3.8 Probability that a call is not completed p_{nc} with $1/\eta = 1/\mu = 3$ min and uniform traffic. ©1999 IEEE. Reprinted, with permission, from [22].

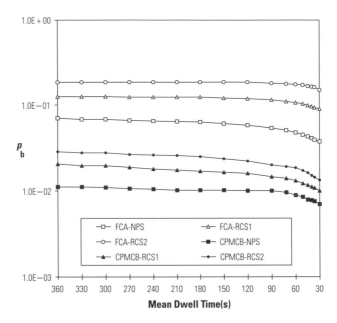

Figure 3.9 Blocking probability p_b with uniform traffic and user mobility effect. Mean dwell time $1/\eta$ is varied by varying η between the range $\eta = 0.5\mu$ to $\eta = 6\mu$. ©1999 IEEE. Reprinted, with permission, from [22].

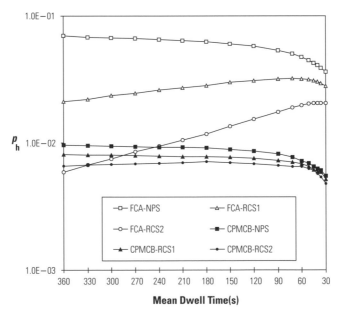

Figure 3.10 Handover failure probability p_h with uniform traffic and user mobility effect. Mean dwell time $1/\eta$ is varied by varying η between the range $\eta = 0.5\mu$ to $\eta = 6\mu$. ©1999 IEEE. Reprinted, with permission, from [22].

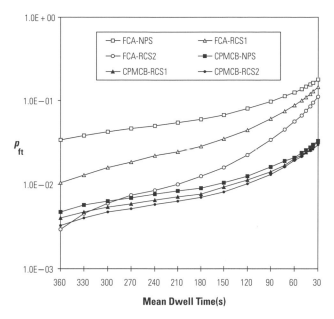

Figure 3.11 Forced termination probability p_{ft} with uniform traffic and user mobility effect. Mean dwell time $1/\eta$ is varied by varying η between the range $\eta = 0.5\mu$ to $\eta = 0.5\mu$. ©1999 IEEE. Reprinted, with permission, from [22].

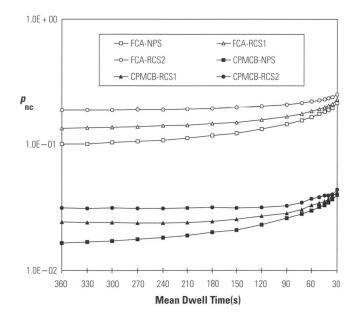

Figure 3.12 Probability that a call is not completed p_{nc} with uniform traffic and user mobility effect. Mean dwell time $1/\eta$ is varied by varying η between the range $\eta = 0.5\mu$ to $\eta = 6\mu$. ©1999 IEEE. Reprinted, with permission, from [22].

traffic loads, but the improvement is less with high traffic loads. Results for p_{nc} are different for the same CAS when a different prioritization strategy is used; but in the case of CPMCB, this is mostly due to the values obtained for p_b.

The general conclusion is that, for uniform traffic mobility, CPMCB, as an overall operating strategy, improves call success appreciably over a wide range of call loadings and mobile terminal mobilities.

3.6.1.2 Uniform Traffic with Fixed Uniform Load; Varying Mobility

Mobility is measured by dwell time (mean $1/\mu$)—the shorter the dwell time the faster the mobile is moving.

For a constant new call arrival rate of 6.8 Erl/cell in the set of Figures 3.9 to 3.12, the mean dwell time is varied by varying η from $\eta = 0.5\mu$ to $\eta = 6\mu$ to assess user mobility effects on p_b, p_h, p_{ft}, and p_{nc}. The values selected for η represent a range variation in the mean dwell time from values as large as 6 min (i.e., 360 sec) and as small as 0.5 min (i.e., 30 sec), respectively. Again, for all four probabilities CPMCB strategies performed better than FCA strategies. In particular,

1. Figure 3.9 shows that when user mobility increases (η increases and cell dwell time decreases) the blocking probability for all the combinations studied decreases with a more rapid fall off at higher mobilities (smaller dwell time). This last effect is similar for both FCA and CPMCB with NPS. However, when RCS is used the increased slope is greater for CPMCB, which probably reflects the greater effect on p_b as mobility increases. As in the case of constant mobility, CPMCB-NPS gives the smallest p_b.

2. Figure 3.10 shows what happens to p_h as the mean dwell time decreases. When the number of handover calls in a cell increases, p_h for CPMCB decreases as it does for FCA-NPS but this is not the case for either FCA-RCS1 or FCA-RCS2. A slight reduction in p_h for these two strategies only occurs for smaller values of cell dwell time. CPMCB strategies present a slight variation in p_h for the cell dwell time range of 360 sec to 120 sec, after which an obvious decrease in this probability is observed.

3. Figure 3.11 shows the variation of p_{ft} versus the mean cell dwell time. From this graph it is clear that p_{ft} increases at a lower rate for CPMCB as user mobility increases and there is little difference between NPS or RCS unlike the situation for FCA. By taking a look at the curves that

represent the FCA-handover strategy, it is easy to tell which combination is the most suitable for a particular type of user according to its speed, but the differences between CPMCB strategies are not so great as to force the choice of one or other.

4. As for p_{nc} in Figure 3.12 it is observed that the combinations of NPS with either access strategy perform better than the other combinations studied, as expected. The curves shown for CPMCB and FCA are similar but with much better p_{nc} values for CPMCB.

3.6.2 Nonuniform Traffic

This section presents results similar to those in Section 3.6.1 but for nonuniform traffic. The actual traffic pattern used is the simplified real one in Figure 3.4 varied in steps from a base load of 0.1 times those shown in the figure. This time both simulation and analysis are used to see how the probabilities p_b, p_h, p_{ft}, and p_{nc} vary with load offered to the cells and with user mobility. Again both FCA and CPMCB CASs are used with the handover schemes NPS and RCS. Results of both simulation and analysis are presented graphically in Figures 3.13 to 3.20.

3.6.2.1 Nonuniform Traffic with Constant Nonuniform Mobility; Varying Asymmetric Load

Figures 3.13 to 3.16 show the performance of the evaluated strategies with an asymmetric traffic load when $1/\mu = 1/\eta = 3$ min. In all these sets of curves, simulated results present a good match to the analytical results, although both are shown. Again, in general, CPMCB performs better than FCA, although this is only clearly shown for p_b and p_{nc}. For p_h and p_{ft} this is only clear for the NPS strategy.

1. As shown in Figure 3.12, CPMCB performs better in terms of p_b but with diminishing improvement as traffic load increases and, if combined with NPS, gives the lowest p_b of all the combinations considered. This time, CPMCB-NPS increases the system capacity only by a factor of 1.57 in comparison with FCA-NPS. In Figures 3.14 and 3.15 it can be noted that CPMCB only performs better than FCA-NPS and FCA-RCS1 for p_h and p_{ft} at values below 0.01.

2. Figure 3.14 shows an important adaptive element given to CPMCB: the self-adaptive characteristic to handover arrivals. It can be observed that CPMCB-RCS2 is the strategy that gives the lowest p_h for the

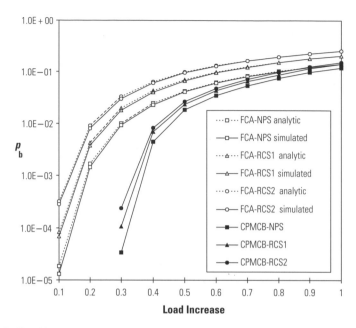

Figure 3.13 Blocking probability p_b with $1/\eta = 1/\mu = 3$ min and nonuniform traffic. ©1999 IEEE. Reprinted, with permission, from [22].

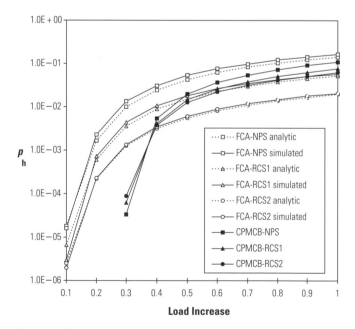

Figure 3.14 Handover failure probability p_h with $1/\eta = 1/\mu = 3$ min and nonuniform traffic. ©1999 IEEE. Reprinted, with permission, from [22].

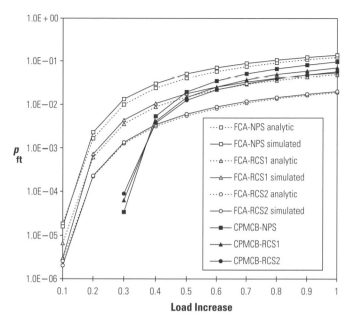

Figure 3.15 Forced termination probability p_{ft} with $1/\eta = 1/\mu = 3$ min and nonuniform traffic. ©1999 IEEE. Reprinted, with permission, from [22].

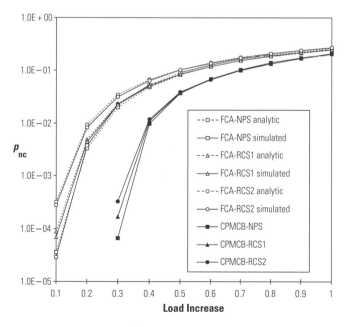

Figure 3.16 Probability that a call is not completed p_{nc} with $1/\eta = 1/\mu = 3$ min and nonuniform traffic. ©1999 IEEE. Reprinted, with permission, from [22].

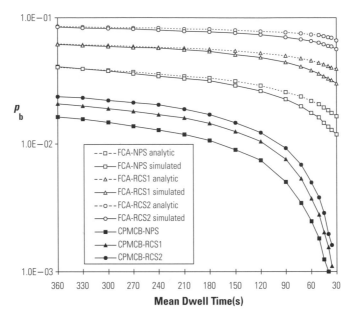

Figure 3.17 Blocking probability p_b with nonuniform traffic and user mobility effect. Mean dwell time $1/\eta$ is varied by varying η between the range $\eta = 0.5\mu$ to $\eta = 6\mu$. ©1999 IEEE. Reprinted, with permission, from [22].

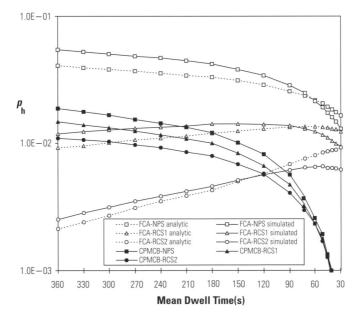

Figure 3.18 Handover failure probability p_h with nonuniform traffic and user mobility effect. Mean dwell time $1/\eta$ is varied by varying η between the range $\eta = 0.5\mu$ to $\eta = 6\mu$. ©1999 IEEE. Reprinted, with permission, from [22].

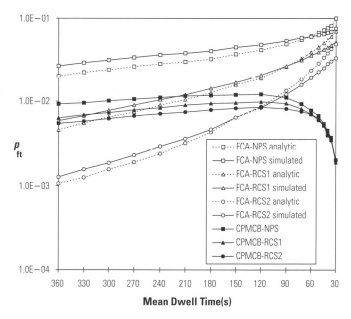

Figure 3.19 Forced termination probability p_{ft} with nonuniform traffic and user mobility effect. Mean dwell time $1/\eta$ is varied by varying η between the range $\eta = 0.5\mu$ to $\eta = 6\mu$. ©1999 IEEE. Reprinted, with permission, from [22].

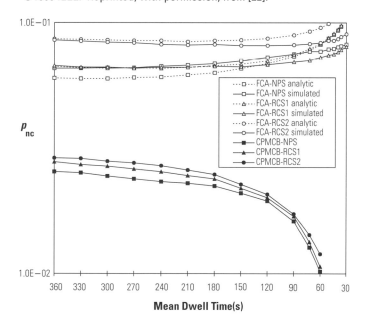

Figure 3.20 Probability that a call is not completed p_{nc} with nonuniform traffic and user mobility effect. Mean dwell time $1/\eta$ is varied by varying η between the range $\eta = 0.5\mu$ to $\eta = 6\mu$. ©1999 IEEE. Reprinted, with permission, from [22].

whole evaluation range, although the differences in capacity increase in comparison with CPMCB-NPS are not significant at $p_h = 0.01$ (the greatest factor found is 1.09). In fact, the close performances of the CPMCB combinations was expected since CPMCB has a cost function that searches for the best balance between new calls and handover arrivals to assign a channel to a given compact pattern and any modification (i.e., giving more priority to handover arrival calls than necessary) will not alter its behavior significantly. This effect is more noticeable when uniform traffic is considered since the number of channels in each cell is not different. This effect can also be seen in Figure 3.15 and in the curves for uniform traffic. As in the case of uniform traffic, Figures 3.14 and 3.15 show that giving priority to handover calls has a greater impact on FCA than on CPMCB. The same grade of service of 0.01 for p_{nc} can be met by CPMCB with a factor of 1.53 times the corresponding load for FCA, according to the results in Figure 3.16. With CPMCB, a small p_h, and therefore a small p_{fh}, do not necessarily mean an increase in p_b and p_{nc}.

3.6.2.2 Nonuniform Traffic with Constant Asymmetric Load; Varying Nonuniform Mobility

Figures 3.17 to 3.20 show how p_b, p_h, p_{fh}, and p_{nc} vary with user mobility (dwell time) in a nonuniform traffic pattern. These results use a constant loading of 4.7 times the nonuniform base load shown in Figure 3.4. These performance values are shown in exactly the same way as was done for the uniform traffic case in Section 3.6.1.2.

Again the overall impression is that CPMCB performs better than FCA. This is clearly so with regard to p_b and p_{nc}, but for p_h and p_{fh} it is only so with NPS priority.

Another obvious point is that disagreement between analytical and simulation results is more apparent than in the case of constant mobility, although still within an acceptable range. The reason they are not as close to each other as in the case of uniform traffic (where analytical and simulations results were in total agreement) is the total offered load seen in simulation with the mobility effect present with nonuniform traffic is slightly different from the total offered load to the same cell when calculated analytically. This occurs when the blocking rate in the system is low, handover arrivals can be approximated by a Poisson process with parameter λ_{hi}, but when the blocking rate is high, handover arrivals do not follow a Poisson process because they are made up of outputs of multiserver loss systems (microcells).

Another reason for this difference between simulated and analytical values is due to the fact that the simulator cannot ensure that the initial distribution of users is the same as the final distribution at the end of each simulation, particularly for the cases of very small dwell times. In fact, the final distribution of users in the simulations is more nonuniform than the initial one, for this particular case. The more uniform the traffic pattern is, the more the simulation results adjust to the numerical results.

Yet another reason for disagreement between results obtained by the two methods is that the central cluster in the nonuniform cellular system used (Figure 3.3) experiences a very high teletraffic demand with the FCA strategy. For this reason, the following comments for FCA are based on numerical rather than simulated results. These sets of values confirm that CPMCB adapts itself better than FCA to user mobility, showing a decrease in all the probabilities as user mobility increases. When user mobility increases, the number of handover arrivals to each cell increases but the occupancy times in each cell decrease. This effect will (as explained in Section 3.3.4) decrease p_b (Figure 3.17) and increase p_{ff} (Figure 3.19) for FCA, both with NPS and RCS. However, p_b with FCA-NPS has a different behavior than with FCA-RCS. With the first combination, p_b decreases as the mean dwell time decreases in all the evaluation range, whereas with FCA-RCS p_b increases as the mobility increases (Figure 3.18). This apparently contradictory performance of FCA-NPS is closely tied to the rate at which p_b decreases in Figure 3.17 as the users move faster. The faster the users are, the faster they leave the cells and the faster they release the resources they are occupying. On average, the rate at which resources will be released (and therefore be available in every cell for both new call and handover arrivals) is faster using NPS rather than RCS. However, FCA-NPS is unable successfully to keep successive handovers of the same call when mobility increases (see Figure 3.19). In p_{nc}, as defined in (20), p_b is weighted by the factor (η/μ). In this way, p_b has more weight than p_b the faster the users are.

Figure 3.20 shows p_{nc} for FCA combined with NPS and RCS. Although it could be argued that by increasing user mobility p_b and p_{ff} should also increase dramatically with CPMCB, exactly the opposite occurs. That happens because CPMCB has a cost function that accounts not only for the new call arrival rate but also for handover call arrivals, making the best possible balance between the two to decrease both p_b and p_{ff}. In CPMCB, channels are assigned to each cell according to the sum of handover and new call arrival rates weighted, respectively, by the expected handover failure and blocking probabilities for every individual compact pattern tried. The reduction in p_b for CPMCB is even larger than with FCA.

In general, these performance values show CPMCB superior to FCA as a CAS. However, this is not so for all assignment/prioritization strategy combinations. Also, where simulated values disagree with analytical ones, the latter are regarded as more reliable and are used in discussion of results.

3.7 Summary

In this chapter, an analytical method is presented to derive the blocking probabilities experienced by new calls (p_b) and handover arrivals (p_h) in a hexagonal-shaped cell belonging to an arbitrary network and for the network as a whole, under uniform and nonuniform traffic distributions. Also, expressions for the forced termination probability (p_{ft}), and the probability that a call is not completed (p_{nc}), were derived under all traffic conditions. The derivation of the sets of probabilities of interest includes the successful derivation of a mathematical expression to calculate the handover arrival rate to each cell and an iterative model with convergence to find this rate, starting with the knowledge of the new call arrival rate only.

Based on the analytical model, an enhancement to the nonuniform compact pattern allocation algorithm was proposed. With this enhancement, the algorithm takes into account the handover arrival rate in its cost functions and decreases not only the blocking probability for new calls but also the handover failure probability. The modified nonuniform compact pattern allocation algorithm was applied to a DCA strategy called CPMCB. This strategy has been shown to enhance system performance under both uniform and nonuniform teletraffic demands.

In the next chapter, the capacity of a hybrid WB-TD/CDMA-TDD and a WB-CDMA-FDD system is assessed in a more complex environment formed of Manhattan-like microcells. FCA strategy is assumed to be used in both cases, but in the case of the WB-TD/CDMA-TDD system a DCA strategy is also used to increase the capacity.

References

[1]   Zang, M., and T. S. Yum, "The Nonuniform Compact Pattern Allocation Algorithm for Cellular Mobile Systems," *IEEE Trans. Vehicular Tech.*, Vol. 40, No.2, May 1991, pp. 387–391.

[2]   Victor, O. K. L., and Xiaoxin Qiu, "Personal Communication Systems (PCS)," *Proc. IEEE*, Vol. 83, No. 9, Sept. 1995, pp. 1210–1243.

[3]   Lin, Yi-bing, et al., "PCS Channel Assignment Strategies for Hand-Off and Initial Access," *IEEE Personal Commun.*, Vol. 1, No. 3, Third Quarter 1994, pp. 47–56.

[4]   Zhang, Ming, and Tak-Shing P. Yum, "Comparisons of Channel Assignment Strategies in Cellular Mobile Telephone Systems," *IEEE Trans. Vehicular Tech.*, Vol. 38, No. 4, Nov. 1989, pp. 211–215.

[5]   Elnoubi, S. M., R. Singh, and S. C. Gupta, "A New Frequency Channel Assignment Algorithm in High Capacity Mobile Communication Systems," *IEEE Trans. Vehicular Tech.*, Vol. VT-31, No. 3, Aug. 1982, pp. 125–131.

[6]   Tekinay, S., and B. Jabbari, "Handover and Channel Assignment in Mobile Cellular Networks," *IEEE Commun. Mag.*, Vol. 29, No. 11, Nov. 1991, pp. 42–46.

[7]   Maric, S. V., et al., "Adaptive Borrowing of Ordered Resources for the Pan-European Mobile Communication (GSM) System," *IEEE Proc. Commun.*, Vol. 141, No. 2, April 1994, pp. 93–97.

[8]   Mihailovic, A., L. Ortigoza-Guerrero, and A. H. Aghvami, "Compact Pattern Based Dynamic Channel Assignment for Reuse Partitioning in Cellular Mobile Systems," in *Proc. Int. Conf. on Telecommun. ICT-98*, pp. 300–304, Porto Carras, Greece, June 21–25, 1998.

[9]   Dragomir, D. Dimitrijevic, and Jelena Vucetic, "Design and Performance Analysis of the Algorithms for Channel Allocation in Cellular Networks," *IEEE Trans. Vehicular Tech.*, Vol. 42, No. 4, Nov. 1993, pp. 526–534.

[10] Katzela, I., and M. Naghshineh, "Channel Assignment Schemes for Cellular Mobile Telecommunication Systems: A Comprehensive Survey," *IEEE Personal Commun. Mag.*, Vol. 3, No. 3, June 1996, pp. 10–31.

[11] Kuek, S. S., "Ordered Dynamic Channel Assignment Scheme with Reassignments in Highway Microcell," *IEEE Trans. Vehicular Tech.*, Vol. 41, No. 3, Aug. 1992, pp. 271–277.

[12] Del Re., E., et al., "Handover and Dynamic Channel Allocation Techniques in Mobile Cellular Networks," *IEEE Trans. Vehicular Tech.*, Vol. 44, No. 2, May 1995, pp. 229–237.

[13] Biaocchi, Andrea, et al., "The Geometric Dynamic Channel Allocation as a Practical Strategy in Mobile Networks with Bursty User Mobility," *IEEE Trans. Vehicular Tech.*, Vol. 44, No. 1, Feb. 1995, pp. 14–23.

[14] Lee, Jongchan, et al., "Channel Allocation and Handover Schemes for Personal Communication Systems," in *Proc. IEEE Vehicular Technol. Conf. VTC'96*, pp. 943–947, Atlanta, GA, 1996.

[15] West, Kevin A., and Gordon L. Stuber, "An Aggressive Dynamic Channel Assignment Strategy for a Microcellular Environment," *IEEE Trans. Vehicular Tech.*, Vol. 43, No. 4, Nov. 1994, pp. 1027–1038.

[16] Madani, K., and A. H. Aghvami, "Investigation of Handover in Distributed Control Channel Allocation (DCCA) for Microcellular Radio Systems," in *Proc. IEEE PIMRC'94*, pp. 160–163, 1994.

[17] Ortigoza-Guerrero, Lauro, and Domingo Lara-Rodríguez, "A Compact Pattern with Maximized Channel Borrowing Strategy for Mobile Cellular Networks," in *Proc. IEEE PIMRC'96*, pp. 329–333, Taipei, Taiwan, Oct. 1996.

[18]   Ortigoza-Guerrero, Lauro, and Domingo Lara-Rodriguez, "A New Dynamic Channel Assignment Strategy for Mobile Cellular Networks based on Compact Patterns with Maximized Channel Borrowing," *IEEE Electronics Letters*, Vol. 32, No. 15, July 1996, pp. 1342–1543.

[19]   Ortigoza-Guerrero, Lauro, and Domingo Lara-Rodríguez, "CPMCB: A Suitable DCA Scheme for the Pan-European GSM System," in *Proc. IEEE Veh. Technol. Conf. VTC'97*, pp. 532–536, Phoenix, AZ, May 1999.

[20]   Bing Lin, Yi, et al., "Queuing Priority Channel Assignment Strategies for PCS Hand-Off and Initial Access," *IEEE Trans. Vehicular Tech.*, Vol. 43, No. 3, Aug. 1994, pp. 704–712.

[21]   Hu, Lon-Rong and Stephen Rappaport, "Personal Communication Systems Using Multiple Hierarchical Cellular Overlays," *IEEE J. Select. Areas Commun.*, Vol. 13, No. 2, Feb. 1995, pp. 406–415.

[22]   Ortigoza-Guerrero, Lauro, and A. H. Aghvami, "Prioritized Handoff Dynamic Channel Allocation Strategy for purchase," *IEEE Trans. Vehicular Tech.*, Vol.48, number 4, July 1999.

[23]   Lin, Yi Bing, Anthony R. Noerpel, and Daniel J. Harasty, "The Sub-Rating Channel Assignment Strategy for PCS Hand-Offs," *IEEE Trans. Vehicular Tech.*, Vol. 45, No. 1, Feb. 1996, pp. 122–129.

4

Capacity Assessment for UMTS

In this chapter a capacity assessment is presented for UMTS by means of system level simulations of the downlink (DL) of two systems to be used in it: hybrid WB-TD/CDMA using asymmetric TDD and WB-CDMA using FDD.

These systems have been deliberately chosen because they correspond, respectively, to the future radio wideband multiple access system (FRAMES) project proposal mode 1 and mode 2 for UMTS Terrestrial Radio Access (UTRA). The cell capacity in Erl/carrier/cell and the spectrum efficiency in Kbps/MHz/cell are assessed for five different circuit switched services in a 5-MHz bandwidth in a microcellular environment with nonuniform traffic. Two different resource allocation strategies are considered in the evaluation of the TD/CDMA-TDD system: the classic FCA and the simple dynamic channel allocation (SDCA) strategies. For WB-CDMA, only FCA is used. Blocking and forced termination probabilities are presented and compared for the TD/CDMA-TDD system using FCA and SDCA.

4.1 UMTS Terrestrial Radio Access

The European research program ACTS (Advanced Communications Technologies and Services) started in the end of 1995 to support collaborative mobile research and development. Within the framework of the ACTS program, the project FRAMES was in charge of defining a multiple access scheme for third generation UMTS.

4.1.1   The FRAMES Project

Within FRAMES, an extensive evaluation study was carried out to investigate different multiple access technologies [1]. Two proposals were made:

1. Wideband direct sequence CDMA (WB-DS-CDMA);
2. WB-TD/CDMA.

The evaluation led to the FRAMES multiple access (FMA) concept, which contains only two modes

1. FMA1, a wideband TDMA system with and without spreading;
2. FMA2, a wideband CDMA system.

These modes were designed to harmonize with each other and to support compatibility with GSM. The FMA concept was presented to the European Telecommunications Standards Institute (ETSI) for consideration in the UMTS standardization process.

4.1.2   ETSI SMG2 Studies

After the FRAMES proposals were made, ETSI SMG2 started a process of selecting one common standard for UTRA in early 1997. Following extensive discussions, the initial proposals were grouped into five concept groups:

1. Wideband DS CDMA: Concept Group Alpha;
2. Wideband TDMA: Concept Group Beta;
3. Orthogonal frequency division multiple access (OFDMA): Concept Group Gamma;
4. Wideband TD/CDMA: Concept Group Delta;
5. Opportunity-driven multiple access (ODMA): Concept Group Epsilon.

Only two of these were selected as possible candidates for UTRA: WB-CDMA and WB-TD/CDMA. The main reason why WB-TDMA was not selected was the unfeasibility of implementing the adaptive equalizer it requires. OFDMA had the drawbacks that there was not a test bed at the time and it needed a very linear amplifier due to its multicarrier nature. ODMA was not a MAS itself. It was not a physical layer concept and was not selected. However, it can be incorporated into WB-CDMA or WB-TD/CDMA.

Two strong groups emerged to study individually each of these two proposals and to improve them in every possible sense. They were the Alpha and the Delta concept groups, in charge of studying, respectively, the DS-WB-CDMA system and the hybrid WB-TD/CDMA system.

SMG2 agreed on a process of selecting the UTRA concept before the end of 1997. According to this process, each concept group presented an evaluation document to the SMG2#23 meeting in October 1997. The evaluation consisted of a description of the concept group's concept and simulations results taking into account predefined models and services.

4.1.2.1 The Alpha Concept Group

Several companies contributed to the Alpha concept. The first inputs to the Alpha concept group were given primarily from FRAMES, Fujitsu, NEC, and Panasonic. With different inputs and different proposals for the WB-CDMA concept, the Alpha concept group went through the merging process to one common WB-CDMA concept. This merging process was finalized at a meeting in Stockholm where all participants agreed on one common WB-CDMA concept in the Alpha concept group.

Having so many companies involved in the Alpha concept group created a working technical discussion with feedback on the proposed solutions from companies with experience from several multiple access techniques. Thus, the merging process toward a common concept resulted in the thoroughly reviewed concept accepted by all participants of the concept group.

The Alpha concept group and the Delta group tried to include the key features for flexible and efficient support of UMTS service needs:

- Support of high data-rate transmissions;
- High service flexibility;
- Good capacity and coverage in the basic system without the need for complex methods;
- Efficient power control;
- Efficient utilization of the achievable frequency diversity with wide-band signal;
- Efficient packet access with a very fast control channel for packet access signaling and packet acknowledgments;
- Spectrum efficient support of HCS;
- No periodicity in the envelope of the uplink transmitted signal avoids problems with audible interference.

4.1.2.2 The Delta Concept Group

The Delta concept group was created after the identification of the two major MAS proposed. The aim of this concept group was to improve the hybrid MAS formed by TDMA and CDMA. The proposal has suffered many changes throughout the evaluation process. The carrier bandwidth and the chip rate are the parameters that have gone through more changes. Originally, the Delta concept group was designed for both FDD and time division duplex (TDD) operation. However, it was decided to choose WB-CDMA for FDD and WB-TD/CDMA for TDD in Paris in 1998. The main change for WB-TD/CDMA (adoption of a 5-MHz broadband carrier and chip rate of 4.096 Mchips/s) was motivated in order to harmonize with the WB-CDMA proposal. It was also motivated to fulfill UMTS requirements for 2 Mbps even in TDD mode and to allow the same RF for FDD and TDD mode. The adoption of a 5-MHz carrier bandwidth has a considerable impact on the overall concept, but the main features contained in the Delta proposal are still maintained.

- A TDMA and a CDMA component, yielding separation of users in the time and code domains;
- Design and use of midambles to allow intracell interference cancellation by means of joint detection;[1]
- No need for soft handover and fast power control due to joint detection receiver design.

The Delta group performed its own evaluations of FMA2 based on link and system level simulations in much the same manner as the Alpha group did for FMA2.

4.1.3 The Decision on MAS for UMTS

Recently, there has been a debate about which MAS should be used in the European UMTS. In the process of defining UTRA, two major options

---

1. Traditionally multi-user detection is divided between interference cancellation (IC) and joint detection (JD). The first category is characterized by the regeneration and subtraction of definite or tentative data estimates. JD means the simultaneous detection of all user's signals, which has to be done anyhow by the BS for UL transmission using the knowledge about intersymbol interference and muliple access interference. In JD the front of the receiver is traditionally (but not necessarily) a bank of correlator filters followed by filters that perform linear or nonlinear transformations, which are usually computationally expensive due to complex matrix calculations and inversions.

have been identified [1–3]: wideband TDMA with spreading feature (WB-TD/CDMA) and WB-CDMA. Both of these proposals meet the requirement of sharing mobile radio channels amongst the largest possible number of authorized subscribers to UMTS services. WB-TD/CDMA and WB-CDMA are based on the proposals for the UMTS radio interface made by the FMA project to special mobile group 2 (SMG) as a support to the ETSI work on UMTS [3].

Strong desires were expressed to find an UTRA proposal with the prospect of becoming a global standard that can encompass the widest possible support. As a result a compromise was reached in ETSI to use WB-CDMA-FDD for paired spectrum and WB-TD/CDMA-TDD for unpaired spectrum. The FDD uplink (UL) (1920 MHz to 1980 MHz) will be located next to the TDD band (1900 MHz to 1920 MHz), which makes FDD/TDD coexistence necessary to some extent [4] (Figure 4.1).

Some controversy has arisen over the use of WB-TD/CDMA-TDD or WB-CDMA in a particular layer of a HCS due to particular characteristics inherent in the technology. WB-TD/CDMA using TDD is said to be the optimum solution to cope with asymmetric traffic generated in microcells due to its duplex technology [5], but nevertheless, these duplex techniques can also be used in WB-CDMA. It has also been said that WB-CDMA will be suitable for operation in both micro- and macro cellular environments. However, none of this has been proven. To identify the best available option, more studies are needed (e.g., to find the system capacity as a function of multimedia traffic, implementation complexity and cost trade-offs, and durability).

4.2 The Purpose of This Chapter

Values for the spectrum efficiency of UTRA are now requested by several groups and for various reasons. They are requested by UTRA forums to refine the evaluation of spectrum requirements for UMTS/IMT-2000 and also in order to evaluate the spectrum to be allocated per operator. In the absence of

Figure 4.1 Frequency allocation for UMTS in Europe.

other references, figures from concept group (Alfa and Delta groups in charge of defining WB-CDMA and WB-TD/CDMA) evaluation documents are sometimes used as a reference for the spectrum efficiency of UTRA, although they have not been fully endorsed by SMG2. It becomes urgent and important for SMG2 to come up with reference models and figures related to the spectrum efficiency of UTRA.

In this chapter, we explore the use of both WB-TD/CDMA-TDD and WB-CDMA-FDD in a microcellular environment of Manhattan-like microcells and compare both systems in terms of system capacity and spectrum efficiency. This is a rather more complicated environment than that in the previous chapter.

4.2.1 WB-TD/CDMA-TDD

FM1 is based on wideband TDMA with and without spreading. The users are separated orthogonally into time slots of different lengths, and within the longest time slots, an additional separation by spreading codes can be used. Thus, FMA1 contains two methods for multiple access: FMA1 without spreading and FMA1 with spreading. Depending on the environment and service, the frame and burst structure can be dynamically adapted. By allocating more or fewer slots and/or spreading codes to a user, bit rates from a few kilobits per second up to 2 Mbps can be provided with good bit-rate granularity [2].

FMA1 has suffered several changes since it was first presented [3]. Originally it was able to support both FDD and TDD. However, it was agreed to choose TDD as the duplexing technique for FMA1 with spreading feature, and it is now called the UTRA-TDD mode. So, WB-TD/CDMA-TDD is based on the FMA1 mode using the spreading feature to increase flexibility and capacity so that this MAS can cope with microcellular environments with asymmetric traffic loads (different loads in the UL and in the DL). The main features of the UTRA-TDD mode are summarized in Table 4.1.

4.2.1.1 Main specifications of UTRA-TDD mode or WB-TD/CDMA-TDD

The introduction of multiple switching points per frame was accepted in the middle of 1998 in one of the versions of the TDD system description. Although the original Delta concept relied on slow (GSM-like) power control, some simulations results showed performance enhancements by faster power control on a per frame basis [5]. In particular, increased performance was achieved in the indoor and pedestrian environments, clearly the scenarios most applicable scenarios for UTRA-TDD. If power control faster than on

Table 4.1
Main Specifications of UTRA-TDD Mode or WB-TD/CDMA-TDD

Feature	Value
Multiple access method	TD/CDMA
Duplexing method	TDD
Channel spacing	5 MHz
Carrier chip rate	4.096 Mchips/s
Time slot structure	16 time slots/frame
Spreading factor	16 chips/symbol
Frame length	10 ms
Multi-rate concept	Variable spreading factor and multi code
FEC codes	Convolutional codes R=1/4 to 1
Modulation	QPSK
Spreading modulation	Linearised GMSK
Channel allocation	Slow and fast DCA supported

a per frame basis is applied, signaling is provided by fast signaling measures. Overall, it seems desirable to allow power control as fast as possible, if more than one time slot is allocated to a service [5].

4.2.2 WB-CDMA-FDD

FMA2 is a wideband DS CDMA system; that is, users are separated by different spreading codes and continuous transmission is used. The basic transmission unit in the resource space is the code. Multiple rates can be achieved through variable spreading factors and multicode options in both UL and DL. Bit rates from a few Kpbs up to 2 Mbps can be provided with good bit-rate granularity. FMA2 has a system chip rate of 4.096 Mchips/s. This allows chip generation from a common clock with FMA1 [1]. The proposed duplexing method for FMA2 is FDD. So far, some work has been done to specify a TDD mode for FMA2, but it has not yet been finished.

As in the case of FMA1, FMA2 has undergone several changes since it was first presented [3]. FMA2 is now referred to as UTRA-FDD mode and will use WB-CDMA-FDD in paired spectrum. The main features of UTRA-FDD mode are summarized in Table 4.2.

Table 4.2
Main Specifications of UTRA-FDD Mode or WB-CDMA-FDD

Feature	Value
Multiple access method	DS-CDMA
Duplexing method	FDD
Channel spacing	5 MHz
Carrier chip rate	4.096 Mchips/s
Interleaving	Intra-frame/Inter-frame
Spreading codes	Short codes, long code optional in UL
Spreading factor	4 to 256 chips/symbol
Frame length	10 ms
Multi-rate concept	Variable spreading factor and multi-code
FEC codes	Convolutional codes R=1/4 to 1
Modulation	DL: QPSK, UL: O-QPSK
Spreading modulation	Linearized GMSK
Channel allocation	Slow and fast DCA supported
Handover	Soft handover

4.3   CASs

Two CASs are used to assess the capacity of a system using WB-TD/CDMA-TDD. They are the classical FCA strategy and the SDCA based on previous proposals [6, 7]. For the WB-CDMA system, only FCA is used.

4.3.1   CASs for WB-CDMA-FDD

For the capacity assessment of the WB-CDMA-FDD system, only the classic FCA strategy is used. In this case, a channel is an available code in the CDMA system in a particular cell. Admission control is used as explained in later sections. The reason why FCA is selected for WB-CDMA is because this MAS consists of two single carriers of 5 MHz, one for the DL and one for the UL. This means WB-CDMA lacks flexibility to use any CAS other than the simple FCA when a maximum bandwidth of 10 MHz is available to cover a single layer. This forces the use of a reuse pattern of 1 because there is only single carrier. WB-CDMA is therefore not suitable for sharing resources between different layers of a HCS.

In case of a layer split in the frequency domain, the available granularity is determined by the carrier bandwidth. The smaller the carrier bandwidth, the better suited a MAS is for deployment of mixed cells. This means that a TDMA- or a OFDMA-based system is particularly well suited for mixed cells, while a WB-CDMA system is not, to mention just the extremes. In the later case, if enough bandwidth is available, a system operator may be able to deploy mixed cell structures but will have to put up with a very inflexible split of resources between layers. In the worst case, an operator may not even have enough bandwidth to deploy more than one layer using FDMA for the layer separation with WB-CDMA air interface. Although a fine frequency granularity may allow an operator to flexibly assign resources to individual layers, static resource partitioning will result in the loss of trunking efficiency. This can be overcome using DCA schemes, which may allow flexible dynamic sharing of resources between layers and thereby adapt to temporal fluctuation of traffic distribution. Therefore, DCA strategies are proposed for WB-TD/CDMA-TDD as discussed in the next section.

4.3.2 CAS for WB-TD/CDMA-TDD

In the evaluation environment of the WB-TD/CDMA-TDD system, two different CASs are considered: classic FCA (previously explained in Chapter 2) and the SDCA strategy [7]. Here, for both strategies, a channel could be a code within a time slot if the spreading feature is used or a simple time slot within the WB-TDMA frame if it is not being used.

The SDCA strategy used is similar to the CAS described in [7]; like CPMCB, it is based on BCO [6]. The main difference between CPMCB and SDCA is that the former is a centralized strategy and the latter one can be applied in a distributed fashion. In this chapter, a channel ordering rather than carrier ordering is used in SDCA. Channel ordering is an ordered list of all available channels. Each cell has its own channel ordering and no two cells separated by less than the reuse distance can use the same channel ordering. Assuming that there are M total channels and there is a reuse factor of N, then each cell will always try to use the first M/N channels in the ordered list. Then, if no available channel is found, a search in the ordered list is performed. The orderings are designed so that the channels occurring at the beginning of a cell's channel ordering occur near the end of a cell's channel ordering in the interference neighborhood. A channel will be acquired if it is not in use in the interference neighborhood. Unlike [6] and as in [7], channels are not divided into nominal and borrowed. When a channel is released in a cell, its place in the ordered list is checked. If it does not correspond to the first M/N channels, then the channel is released; otherwise, a channel reallocation is made to release any possible busy channel appearing at the end of the ordered list. Note that this strategy does not need any exchange of information within the interference neighborhood. The busy/idle status of carriers can be determined by passive nonintrusive monitoring at each BS.

In fact, this is a distributed strategy and is suboptimal since every cell has access only to partial information. However, using distributed computation and the reduced communication needs make it feasible for UMTS use, where the signaling load between BSs is required to be minimum.

4.3.2.1 Additional Management of Resources in TDD Mode

The UTRA TDD mode can be addressed with two main approaches [8]. It can first be considered as some kind of slotted WB-CDMA system, in which users are also separated over the whole frame in each direction and interference (intra- and intercell) is controlled by fast power control and macrodiversity. This approach favors a continuous transmission considering both directions and relies on a variable spreading factor of up to 256 [8].

However, the UTRA-TDD mode can also be considered as a TDMA system with a CDMA component inside its time slots. Intracell interference is reduced/removed by a joint detection procedure, provided the number of codes superposed on one time slot is not too large. The intercell interference can be reduced by a time reuse cluster or even a frequency reuse cluster if several carriers are available. Therefore, fast power control and macrodiversity are not required with this approach. This scheme relies on the good performance of joint detection and this on the maximum number of codes per slot, which should not exceed 10 to 12 [8].

There are several reasons for using the minimum number of slots. Indeed, if multicode transmission is used, the complexity of joint detection only depends on the number of slots being used, whatever the number of codes allocated to the user on each slot. On the other hand, if a variable spreading factor is used to avoid multicode transmission and joint detection, it is also better to allocate all resources on one slot to the same user. In order to reduce the processing power, it is then better to allocate resources on the lowest number of slots [8].

WB-TD/CDMA-TDD relies on a real TDMA component; that is, users are mapped on as few time slots as possible in order to allow DCA algorithms to prevent high interference conditions. This principle combined with the possibility of removing intracell interference by advanced receiver techniques implies small spreading factors. Variable spreading with spreading factors lower than or equal to 16 is proposed in order to maintain the TDMA component of the system [5].

WB-TD/CDMA-TDD has to be able to cover asymmetric services because the flexibility in DL/UL slot allocation allows efficient use of spectrum for asymmetric traffic. In order to maintain as much flexibility as possible and at the same time allow for fast power control whenever useful, the following basic DL/UL slot allocations for symmetric and asymmetric traffic were proposed [5]:

- At least two time slots have to be allocated to DL transmission, from BS to MS, while at least one time slot has to be allocated to UL transmission, from MS to BS, for random access reasons [5].
- DL and UL resources are to be distributed in space first (code pooling) and then in time [9].

The TDD mode, with a 4-Mchip/s chip rate, is being defined for a frequency reuse of one with 14 slots per frame that can be used for traffic (two are reserved for BCCH and RACH). However, for service bit rates above 144 Kbps, assuming a symmetric split of resources between UL and DL, the processing gain becomes lower than 10 dB, so it becomes difficult to implement a network with a reuse pattern of one. Either time or frequency reuse is therefore needed to efficiently support services above 144 Kbps [8]. It should be possible to support around 144 Kbps on one slot, assuming 9 or 10 codes are allocated (or one code with spreading factor of two). In these conditions, this slot becomes too loaded to be reused in adjacent cells and time reuse has to be implemented. The reuse factor will depend on the type of environment but should typically be of the order of two or three. It is clear that with a reuse of one, assuming only six codes per slot could be used, the terminal power consumption would be doubled at the same bit rate.

Of course, this strategy puts some constraints on resource allocation and implies, for instance, the grouping of all speech users in a single slot before allocating a new one in order to keep slots free for high bit rate users. This management of resources may require some kind of reshuffling, which could increase the delay in allocation [8].

The assignment of resources is carried out as follows:

- All speech users are multiplexed on the same slot and separated by joint detection.
- Users with higher bit rates are allocated one complete slot plus several codes (or one code with a spreading factor below 16) on another slot, which is shared, for instance, with other speech users. The joint detection allows codes to be received with different spreading factors.
- When one slot is shared among all cells it is only loaded with six codes in each cell.

4.4   Evaluation Environment

A system-level simulation is done according to the specifications in [10] and [3] for the spread case of FMA1 and FMA2 to determine the cell capacity and the spectrum efficiency for both cases. CIR evaluations were performed on a call-by-call basis to determine whether or not a call will experience the required CIR ratio for the connection.

4.4.1   Microcellular Layout

An environment consisting of a microcell layer formed of a (regular) rectangular grid of intersecting streets (usually referred to as Manhattan-like microcells), as shown in Figure 4.2, was assumed in the simulations. The BSs are at street lamp height and are placed at the center of the crossroads (clover leaf cells). In a real system this would ease the problems related to handover since users do not experience a sudden drop of the desired signal when turning a street corner [11]. However, positioning BSs at crossroads will not prevent calls' signal strengths from falling rapidly. In reality, this is exactly what does happen, especially when the antennas are below roof height (necessary for the microcell structure) unless care is taken to employ appropriate antennas or other techniques to ensure continuity of RF signal round the corners. This BS deployment is said to be the ultimate step in a microcellular network to cope with high traffic loads. BSs are separated by 200m. Users are only tracked at the microcell level, and a wraparound topology is used so that boundary effects are eliminated in the simulations, as depicted in Figure 4.2. The interference neighborhoods of each cell also wrap around. This means that if two cells are on opposite edges of the grid, subscribers leaving one cell to enter the other may not use the same code or time slot simultaneously in both cells.

According to [8], a cluster size in the range of one to three has to be supported for WB-TD/CDMA-TDD. In our simulations, a reuse pattern of 1 was used for both the WB-TD/CDMA-TDD and WB-CDMA-FDD systems. A 5-MHz bandwidth was assumed under an asymmetric TDD approach for the TD/CDMA-TDD, whereas for the WB-CDMA-FDD system, a 10-MHz bandwidth was considered. 5 MHz are used for the UP and 5 MHz for the DL.

4.4.2   User Mobility

The user mobility model is highly related to the Manhattan-like structure defined earlier. With, this environment, mobiles move along the streets and

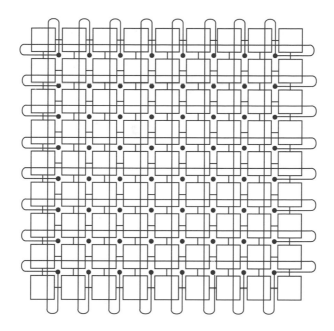

Figure 4.2 Evaluation environment.

may turn at cross-streets with a given probability, as shown in Figure 4.3. The turn probability, *Turn_Prob*, is equal to 0.67 with equal probability for turning left and right (0.33 and 0.33, respectively).

4.4.3 Propagation Model for the Microcellular Layer

For the microcell layer, a three-slope model is adopted [12]. The model is given mathematically by the following expressions for the line of sight (LOS) condition:

$$L_{LOS} = \begin{cases} L_b + 10n_1\log\left(\dfrac{d}{R_b}\right) & d \leq R_b \\[4mm] L_b + 10n_2\log\left(\dfrac{d}{R_b}\right) & d \geq R_b \end{cases} \qquad (4.1)$$

and for the no line of sight condition (NLOS):

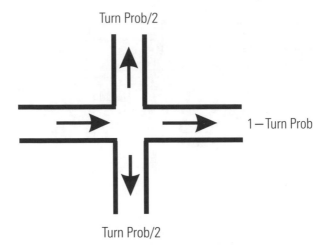

Figure 4.3 Mobility parameters at a crossroad point.

$$L_{NLOS} = L_{LOS}\left(d_{corner}\right) + L_{corner} + 10n_{NLOS}\ \log\!\left(\frac{d}{d_{corner}}\right) \tag{4.2}$$

Each BS is on a street lamp, so the buildings act as waveguides. A breakpoint, located at a distance R_b from the transmitter, marks the separation between the two LOS segments. The second of these LOS segments predicts larger losses than the first one. When there is NLOS, the losses are given by three components:

1. The LOS component (as if the mobile were at a distance d_{corner} from the transmitter);
2. The additional losses caused by turning at a crossroad;
3. The proper NLOS condition.

   In the equations d stands for the distance from the transmitter to the receiver measured along the street path, d_{corner} is the distance from the transmitter to the corner, in the case of NLOS.

   The rest of the parameters involved are

$$R_b = \frac{4\,h_b\,h_m}{\lambda} \tag{4.3}$$

$$L_b = \left| 20 \log_{10}\left(\frac{\lambda^2}{8\pi h_b h_m} \right) \right| \tag{4.4}$$

$$L_{corner} = -0.1w_s + 0.05d_{corner} + 20 \tag{4.5}$$

$$n_{NLOS} = -0.05w_s + 0.02d_{corner} + 4$$

where h_b and h_m are the BS and MS antenna heights, λ is the wavelength, and w_s is the street width. The shadowing effect is modeled as $10^{\varepsilon/10}$, where ε is a Gaussian variable with zero mean and standard deviation of 4 dB. Shadowing is modeled so that the fading process is not correlated between BSs; however, for a more realistic simulation a degree of correlation must exist between them and multipath fading should also be modeled.

The parameters involved in the simulation using this propagation model are presented in Table 4.3. In the simulations there are no reuse constraints on NLOS cochannel cells because of the large signal attenuation around a corner [7].

4.4.4 Teletraffic Model

As already stated, users are only tracked in the microcell layer. This allows the active-dormant Markov model to be used in the simulations to account for even and uneven teletraffic distributions in the scenario proposed. This model was originally presented in [13] and in a modified way to model handover queuing in [7]. Making use of this model, traffic is initially uniform and changes with time according to a random process that modifies the offered load to each cell by a factor that actually depends on the traffic load in the whole

Table 4.3
Summary of the Physical Parameters Used in the Simulation

Parameter	Value
BS antenna height h_b (m)	12
MS antenna height h_m (m)	1.5
Frequency (MHz)	1900
Street width w_s (m)	30
n1	2
n2	4

system. The model is Markovian, so all events occur with exponentially distributed interarrival times. The parameters of the distributions change with time to reflect the time-varying nature of the model. The state of any microcell i, for type of service j, at any time can be described by the following parameters: new call arrival rate ($\lambda i,j$) and number of active calls ($Nactive,i,j$), where $\lambda i,j$ is the rate at which new users of type j arrive at cell i. $Nactive,i,j$ is the number of active users of type j in cell i that are being serviced by cell i.

As the simulation proceeds, five types of event are generated for each type of service: new arrivals, completions, handover attempts, active-to-dormant mode transitions, and dormant-to-active mode transitions. All events occur independently; therefore, five random times are generated for each type of service and the next event corresponds to that with the minimum time. Once an event is selected and the type of user identified, the event must be randomly assigned to a cell according to [7]. Once cell i is selected, a type of service must be associated with the event just picked up. The event selection procedure is fully described in [7]. The parameters used in the simulation are summarized in Table 4.4. Asymmetric traffic was generated assuming that a relation of 3 to 1 exists between the traffic in the DL and the corresponding traffic in the UL.

4.4.5 Types of Service

The capacity assessment is done for a quality of service (QoS) of 1% in terms of the blocking probability of new call arrivals. Two types of circuit-switched service are considered in the system: speech and data. For data transmissions

Table 4.4
Parameters Used in the Simulation

Parameter	Value or Range
Offered traffic load per microcell (Erl)	0-90
Active to dormant traffic ratio	5
Average duration of DORMANT mode (min)	10
Average duration of ACTIVE mode (min)	1
Average number of handover per call	0.28
Average number of codes per time slot	8
Duplex technique (asymmetric) for TD/CDMA	TDD
No. of 5 MHz carriers	1
Total number of microcells	72

Table 4.5
Types of Services Considered ©1999 IEEE. Reprinted, with permission, from [14].

Service	Rate (Kbps)	Type
speech	8	1
data	32	2
data	64	3
data	144	4
data	384	5

four different rates are taken into account, as shown in Table 4.5. The unencumbered session duration of a call is taken as 3 min. The cell capacity and the spectrum efficiency are assessed independently for every type of service in the DL only. Type of service 1 is assumed to have 50% voice activity.

4.4.6 CIR Threshold

Only the DL CIR ratio is calculated since it is assumed that this will be worse than the UL and thus provide the limit to system performance. The CIR is calculated according to the corresponding propagation model and the MAS used to establish a connection: TDMA with or without spreading and CDMA. In the case of WB-TD/CDMA-TDD without spreading, only intercell interference is taken into account (cochannel interference), while for the other cases, inter-and intracell interference have to be considered. Once the CIR is calculated accordingly, this ratio is matched to the appropriate Eb/N_0, as specified in Table 4.6.

The relation between the average signal-to-noise ratio (SNR) Eb/N_0 (Eb is the energy per bit and N_0 is the noise power spectral density) and the CIR, with I denoting the interference power, is given by [15]

$$\frac{C}{I} = \frac{Eb}{N_0}\frac{R_c \log_2 M}{BQT_c} \tag{4.6}$$

where R_c is the rate of the channel encoder (depending on the service), M is the size of the data symbol alphabet (4 for QPSK, 16 for 16 QAM), B is the user bandwidth (5 MHz), Q is the number of chips per symbol, and T_c is the chip duration.

Table 4.6
Required Eb/N_0 for Every Service Considered in the Evaluation

System	Environment	Service Type	Eb/N₀ in the DL
WB-TD/CDMA-TDD	Pedestrian	1	6.1 dB @ 10^{-3} [16]
	(microcells)	2	3.3 @ 10^{-6} [16]
		3	3.1 @ 10^{-6} [16]
		4	2.1 @ 10^{-6} [16]
		5	1.4 @ 10^{-6} [16]
WB-CDMA-FDD	Pedestrian	1	8dB @10^{-3} [17]
	(microcells)	2	6dB @ 10^{-6}
		3	5dB @ 10^{-6}
		4	2dB @10^{-6} [17]
		5	3.2 @ 10^{-6} [17]

4.4.7   Additional Information on the Simulation of the WB-CDMA System

A fast signal-to-interference ratio (SIR) is assumed in the DL and the power of the transmitters is balanced to meet the average SIR during one frame. The DL power control may introduce a near–far problem if a user near the BS is interfered by the power transmitted to a user at the cell border due to a nonorthogonal DL. This problem is avoided by having a limited dynamic range in the DL. A 20-dB dynamic range per bearer (traffic channel) is assumed in the simulations. Soft handover is used for all the services (circuit switched). The soft handover algorithm simply connects the strongest BSs within the handover window based on path loss (excluding fast fading). For the DL, rate matching to the closest bit rate, using either unequal repetition or code puncturing,[2] is done for all the services. A single code option is assumed in the evaluation.

---

2. Typical convolutional codes intrinsically have code rates expressed as simple fractions like $^1/_2$ or $^1/_3$. When a code having higher rate (less redundancy) is needed, one of these lower-rate "mother" codes is punctured, that is, some of the coded bits are just not transmitted, according to a regular pattern known to the receiver. At the receiver "dummy bits" are reinserted to replace those omitted but are marked as erasures—bits having zero confidence—so that the Viterbi decoder treats them accordingly. Punctured codes obviously are less powerful than the mother code, but there is an acceptable steady trade-off between performance and code rate as the degree of puncturing is increased.

4.4.8 Additional Information on the Simulation of the WB-TD/CDMA-TDD System

Since the hybrid system TD/CDMA-TDD is based on a TDMA frame structure, the basic way of supporting multiple bit rates is to assign multiple slots to a user. For operation with spreading, multiple codes on a slot can also be assigned. The bit rate for a specific service is fine tuned by selecting an appropriate combination of slot length, burst type, number of slots, and coding rate [3]. This is the approach we assume in the simulations. The number of required codes per time slot per user for every service considered is shown in Table 4.7.

4.5. Discussion of Results

The main aim of this chapter is to show the reader a comparison of the two different systems proposed for UMTS in terms of cell capacity and spectrum efficiency. These concepts are tightly tied and may mean different things for a user and for an operator.

On the one hand, from a customer point of view, a system with high spectrum efficiency—or a system with high cell capacity—would represent a system with reliable service that may be translated into few disruptions, few calls terminated abruptly, and service availability, for example. On the other hand, from an operator point of view, a system with high spectrum efficiency would offer a relatively high QoS and/or an increase in the number of users that can possibly be attended and therefore an increase in the income originated. Alternately, if the QoS intended to be provided is lower than that actually being offered, network operators can optionally decide to decrease

Table 4.7
Number of Codes Required for Every Service in the WB-TD/CDMA-TDD System

Service Type	Number of Codes Required [16]
1	1
2	2
3	4
4	9
5	27

the QoS to the value required and decrease the running cost by decreasing the infrastructure.

Much has been said about the advantages and disadvantages inherent to UMTS modes. The decision to select one or another mode certainly is not only limited to the spectral efficiency that each of them can offer but is influenced by several other factors. The simulations in this chapter contribute in a small way to the process of selecting the best UTRA mode for a particular scenario (indoor–outdoor) and for particular types of service (circuit-switched) by comparing them in terms of spectrum efficiency and cell capacity.

Bearing this idea in mind, the comparison of blocking and forced termination probabilities, obtained using FCA and SDCA in a WB-TD/CDMA-TDD system for all the five types of service considered under noncellular network uniform traffic conditions, are presented in this section. Also, the cell capacity and the spectrum efficiency for WB-TD/CDMA-TDD and WB-CDMA are obtained in the simplified system described in Section 4.3.1.

4.5.1 Blocking and Forced Termination Probabilities for WB-TD/CDMA-TDD

The set of Figures 4.4, 4.6, 4.8, 4.10, and 4.12 shows, respectively, the overall blocking probability versus the offered load per microcell for types of service 1, 2, 3, 4, and 5 when using FCA and SDCA strategies in a 5-MHz bandwidth in a WB-TD/CDMA-TDD system. In each case, the blocking probability is plotted versus the offered load per microcell. The graphs shown are applicable to the DL only.

As can be observed from the curves for the overall blocking probability and for the forced termination probability, the SDCA strategy always performs better than the FCA strategy. A blocking probability of 1% is always reached faster at lower loads with the FCA strategy than with the SDCA, therefore producing a smaller cell capacity. The cell capacity increase in percentage can be easily calculated from the figures in Table 4.8, where the cell capacity is reported. Using SDCA rather than FCA considerably decreases both the overall blocking and the forced termination probability at light and medium traffic loads for each case, with diminishing advantages as the load increases.

The set of Figures 4.5, 4.7, 4.9, 4.11, and 4.13 shows the forced termination probability versus the offered load per microcell for the cases being studied. Using SDCA instead of FCA in the system always reduces the forced terminated calls; however, as the number of channels required for data users increases, the advantages of SDCA compared with FCA are less noticeable.

Traditionally, as it was shown in Chapter 3, the use of a DCA strategy in a cellular system using any MAS with a FDD approach can be translated

Table 4.8
Cell Capacity in a 5-MHz Bandwidth @1999 IEEE. Reprinted, with permission, from [14].

Service Type	Cell Capacity (Erl/carrier/cell)		
	WB-CDMA	WB-TD/CDMA	
	FCA	FCA	SDCA
1	56.25	57	85
2	34	27.5	38
3	16.5	12.5	18.5
4	5.5	4.2	6.8
5	2.3	0.8	2.12

in an increase of the offered traffic being carried or in a reduction in the over-all blocking probability. However, for the particular case of a system like WB-TD/CDMA that uses TDD, the use of SDCA also implies an improve-ment in the way the carrier is being shared between UL and DL, as shown in

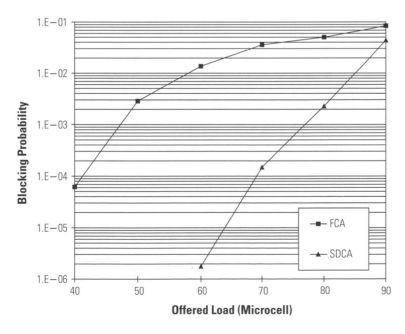

Figure 4.4 Blocking probability for type of service 1. ©1999 IEEE. Reprinted, with permission, from [14].

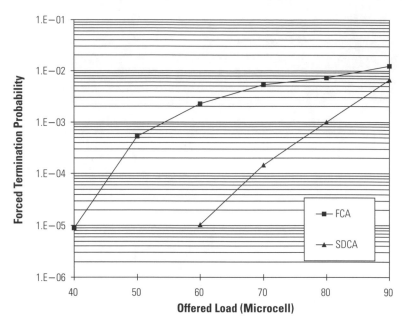

Figure 4.5 Forced termination probability for type of service 1. ©1999 IEEE. Reprinted, with permission, from [14].

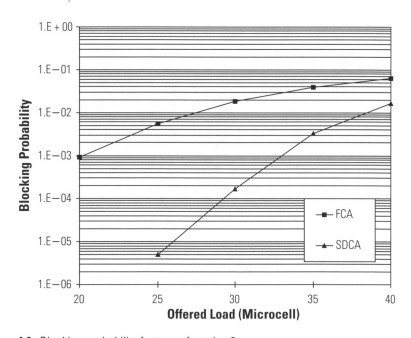

Figure 4.6 Blocking probability for type of service 2.

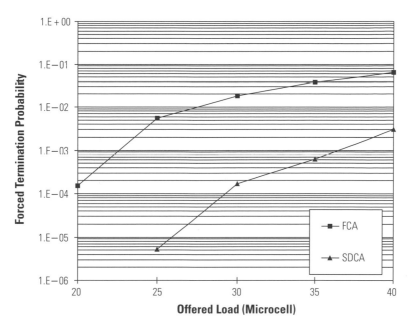

Figure 4.7 Forced termination probability for type of service 2.

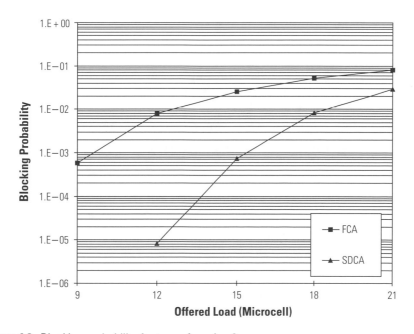

Figure 4.8 Blocking probability for type of service 3.

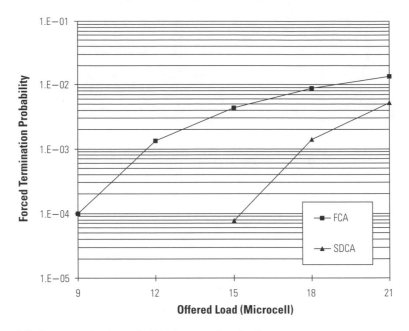

Figure 4.9 Forced termination probability for type of service 3.

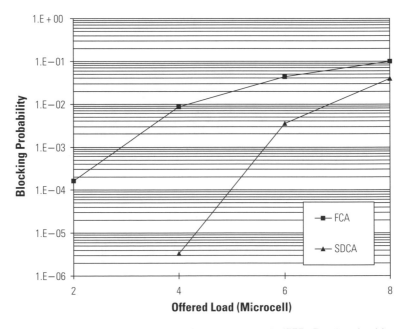

Figure 4.10 Blocking probability for type of service 4. ©1999 IEEE. Reprinted, with permission, from [14].

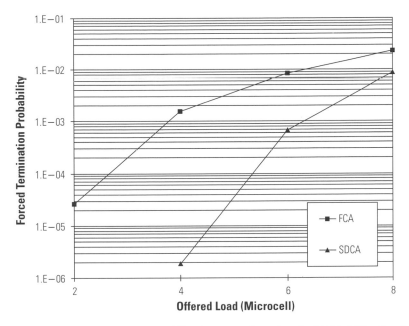

Figure 4.11 Forced termination probability for type of service 4. ©1999 IEEE. Reprinted, with permission, from [14].

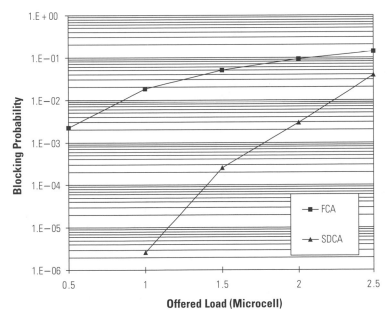

Figure 4.12 Blocking probability for type of service 5. ©1999 IEEE. Reprinted, with permission, from [14].

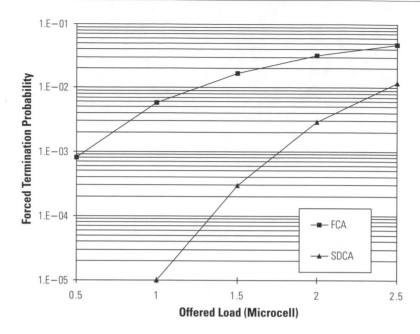

Figure 4.13 Forced termination probability for type of service 5. ©1999 IEEE. Reprinted, with permission, from [14].

the previous sets of graphs. This is accomplished by reallocations of codes in different time slots, as explained in Section 4.2.2.1. Reallocations of calls attended by several codes in different time slots can release time slots than can be assigned for the UL or DL by varying the switching point to attend an asymmetric connection when required.

4.5.2 Capacity Assessment

Tables 4.8 and 4.9 show, respectively, the cell capacity and the spectrum efficiency evaluation for the WB-TD/CDMA-TDD and WB-CDMA systems.

In Table 4.8, cell capacity is expressed in Erl/carrier/cell and obtained when the blocking probability meets an overall value of 1%. From the results inTable 4.8, it can be observed that SDCA increases the cell capacity compared to FCA in the case of WB-TD/CDMA-TDD. It can also be observed that WB-TD/CDMA-TDD combined with a SDCA strategy can improve the cell capacity to such an extent that it is even larger than that offered by WB-CDMA for all types of services but one. WB-CDMA only performs marginally better for service type 5.

Table 4.9
Spectrum Efficiency in a 5-MHz Bandwidth. ©1999 IEEE. Reprinted, with permission, from [14].

Service Type	Spectrum Efficiency (kbps/MHz/cell)		
	WB-CDMA	WB-TD/CDMA	
	FCA	FCA	SDCA
1	45	45.6	68
2	217.6	176	288.8
3	211.2	160	236.8
4	158.4	120.9	195.8
5	176.6	61.4	162.8

Since the results in Table 4.9 are directly calculated from the results in Table 4.8, the same reasoning as above can be applied to the spectrum efficiency case. For service types 1 to 4, the combination of SDCA and WB-TD/CDMA-TDD presents the largest spectrum efficiency, whereas for service type 5, WB-CDMA offers the best performance.

It is worth pointing out that because a reuse pattern of one was used in the WB-TD/CDMA-TDD system, assuming only six codes/slot, the terminal power consumption would be doubled to support the same bit rate, as in the case with reuse patterns different from one.

Results in Tables 4.8 and 4.9 lead to the following conclusions.

- When a reuse pattern of one is used for both the WB-TD/CDMA and the WB-CDMA systems and the FCA is used as the CAS, the UTRA-FDD mode actually performs better than the UTRA-TDD mode. A higher spectral efficiency is shown for UTRA-FDD for most of the services and is only comparable for service type 1, which is voice. The performance of UTRA-FDD mode is much better the higher the bit rate of the communication is. However, it should always be remembered that this in part is because the UTRA-TDD mode only uses half of the bandwidth assigned to UTRA-FDD.

- When the UTRA-TDD mode makes use of a SDCA, an enormous increase in the spectrum efficiency is experienced by this mode when compared to that obtained when FCA is used. With SDCA,

UTRA-TDD performance is comparable to UTRA-FDD's performance and, in fact, are slightly better for the four slowest services.

- Although the evaluation of spectrum efficiency is performed in a 5-MHz bandwidth, the process is not at all fair for the UTRA-TDD mode. The reader must remember that this system is using a single 5-MHz carrier to carry traffic in the UL and DL whereas the UTRA-FDD has dedicated 5-MHz carriers for the UL and DL traffic. It is under these circumstances that it can be concluded WB-TD/CDMA-TDD is much more efficient in terms of spectrum utilization than WB-CDMA. However, the use of a DCA strategy brings the disadvantages described in Chapter 2.

- The combination of a DCA strategy with UTRA-TDD mode helps this mode to face asymmetric traffic conditions. By examining Tables 4.8 and 4.9, it is highly advisable to use the UTRA-TDD mode for microcellular environments, where, presumably, highly asymmetric traffic will be offered.

4.6 Summary

The topics of cell capacity, spectrum efficiency evaluation, and resource allocation for WB-TD/CDMA-TDD and WB-CDMA systems have been analyzed and simulated for a simplified Manhattan-like microcellular system. For the case of WB-TD/CDMA-TDD, it has been shown that using a DCA strategy rather than the classic FCA means improved efficiency in the task of channel managing, even when the DCA strategy is very simple. Speech and data transmission services at 32 Kbps, 64 Kbps, and 144 Kbps can be supported by a WB-TD/CDMA system combined with an SDCA strategy meeting the required QoS with better performance than the WB-CDMA strategy.

References

[1]   Ovesjo, Frederik, et al., "Frames Multiple Access Mode 2- Wideband CDMA," in *Proc. IEEE, PIMRC'97*, pp. 42–46, Helsinki, Finland, Sept. 1997.

[2]   Klein, Anja, et al. "FRAMES Multiple Access Mode 1—Wideband TDMA With and Without Spreading," in *Proc. IEEE PIMRC'97*, pp. 37–41, Helsinki, Finland, Sept. 1997.

[3]   ETSI STC SMG2#20, *FRAMES Multiple Access Scheme Proposal for the UMTS Radio Interface-SMG2 Workshop on UMTS Radio Interface Technologies*, Sophia Antiopolis, Dec. 16–20, 1996.

[4] ETSI SMG2 UMTS Ad-hoc, Tdoc SMG2 UMTS 137/98, "The Use of FDD and TDD Modes in UTRA," Stratford, United Kingdom, July 6–8, 1998.

[5] ETSI STC SMG2 UMTS L1 #6, Tdoc SMG2 305/98, "UTRA TDD Harmonisation Proposal," Helsinki, Finland, Sept. 8–11, 1998.

[6] Elnoubi, S. M., R. Singh, and S. C. Gupta, "A New Frequency Channel Assignment Algorithm in High Capacity Mobile Communication Systems," *IEEE Trans. Vehicular Tech.*, Vol. VT-31, No. 3, Aug. 1982, pp. 125–131.

[7] West, Kevin A., and Gordon L. Stuber, "An Aggressive Dynamic Channel Assignment Strategy for a Microcellular Environment," *IEEE Trans. Vehicular Tech.*, Vol. 43, No. 4, Nov. 1994, pp. 1027–1038.

[8] ETSI SMG2 UMTS L1 Expert Group #5, Tdoc SMG2 UMTS L1 263/98, "Radio Resource Management Strategies for the UTRA TDD Mode and Implications on the Spreading Factor," Gatwick, United Kingdom, July 15–17, 1998.

[9] ETSI SMG2 UMTS L1, Tdoc SMG2 UMTS-L1 328/98, "Resource Allocation in Joint Predistortion Systems," Helsinki, Finland, Sept. 8–11, 1998.

[10] UMTS 30.03, V3.0.0 (1997-05), DTR/SMG-50402 (3PC0032S.pdf), ETSI, SMG-5, "Universal Mobile Telecommunication System (UMTS); Selection Procedures for the Choice of Radio Transmission Technologies of the UMTS," Sophia Antipolis Cedex, France, 1997.

[11] Frullone, M., G. Riva, P. Grazioso, and C. Carciofi, "Analysis of Optimum Resource Management Strategies in Layered Cellular Structures," in *Proc. Int. Conf. on Universal and Personal Commun., ICUPC'94*, pp. 371–375, San Diego, CA, USA. 1994.

[12] Pizzarroso, M., and J. Jimenez, "Common Basis for Evaluation of a TDMA and CODIT System Concepts," R2020/TDE/CA/DS/L/SIG5-1/al,25/08/95.

[13] Nanda, S., and D. J. Goodman, "Dynamic Resource Acquisition: Distributed Carrier Allocation for TDMA Cellular Systems," *Third Generation Wireless Information Networks*, New York: Kluwer, pp. 92–124, 1994.

[14] Ortigoza-Guerrero, Lauro, and A.H. Aghvami, "Capacity Assesment for UTRA," *Proceedings of the IEEE Vehicular Tech. Conference,* VTC99 summer, pp. 1653-1657, Houston, TX, 1999.

[15] Blanz, Josef, et al., "Performance of a Cellular Hybrid C/TDMA Mobile Radio System Applying Joint Detection and Coherent Receiver Antenna Diversity," *IEEE Trans. Vehicular Tech.*, Vol. VT-12, No. 4, May 1982, pp. 125–131.

[16] ETSI STC SMG2 UMTS L1#6, Tdoc SMG2 306/98, "UTRA TDD Link Level and System Level Simulation Results for ITU Submission," Helsinki, Finland, Sept. 8–11, 1998.

[17] ETSI SMG2#23, Tdoc SMG2 270/97, "Concept Group Alpha–Wideband Direct Sequence CDMA, Evaluation Document (Draft 1.0), Part 1: System Description, Performance Evaluation," Bad Salzdetfurth, Germany, Oct. 1–3, 1997.

5

Spectrum Partitioning in a Microcell/Macrocell Cellular Layout with Overflow

5.1 Introduction

When there is a single bandwidth to be shared between the microcell and the macrocell layers in a mobile cellular telecommunications system with a HCS, spectrum partitioning becomes an important issue. The optimal amount of bandwidth has to be given to every layer so that the overall blocking in the HCS is at a minimum.

Spectrum partitioning is a function of several parameters but depends primarily on the speed of users in the microcells and macrocells and on the number of microcells overlaid by a single macrocell.

This chapter contains a teletraffic analysis to model a hierarchical cellular structure formed of two layers and a dynamic method to divide the spectrum. Both the teletraffic analysis and the spectrum partitioning technique are based on more general methods that have previously been presented individually in the literature. They are combined, after some modifications, to perform the partitioning.

Although in the rest of this book a grade of service (i.e., blocking probability) of 1% is always ensured in the evaluation of system performances, in this chapter the spectrum partitioning is calculated using the criterion that the blocking probability in each layer (macrocellular and microcellular), when allowing overflow traffic from microcells to macrocells, ensures a grade of service of 2%. This is done for a reason. At the end of this chapter a comparison is made between the results obtained with the methodology given here and those previously obtained by an older method of spectrum partitioning that used a criterion of 2% in terms of blocking probability. FCA is used in both levels of the hierarchy.

The teletraffic analysis is based on a multidimensional birth and death process, and the calculation assumes perfect assignment of mobiles to each of the layers according to their speeds (i.e., the probability of incorrect assignment of a call is zero). Users' mobility is taken into account in the model by changing the cell dwell time in every cell in each layer of the hierarchy according to the mean speed of users in the level they are being attended. The offered load to each layer depends on the probability density distribution for the speed of mobiles in the system.

In addition, the effect of reserving channels for handovers in every layer of the hierarchy is assessed using the RCS and the NPS introduced in Chapter 2. Several possible combinations are tried in the microcells and macrocells to find out which one produces the best results in terms of traffic-carrying capacity. The combinations studied are:

- NPS in both microcell and macrocell layers;
- RCS in both microcell and macrocell layers;
- RCS in the microcell layer, and NPS in the macrocell layer;
- NPS in the microcell layer, and RCS in the macrocell layer.

The technique presented here for spectrum partitioning can be dynamically implemented by changing the ratio of the channels allocated to the microcells and their overlying macrocell until an optimum partitioning is found, which carries the maximum offered load.

5.2 The HCS Scenario

In a macrocell/microcell cellular system, macrocells are used to cover a wide area while microcells are used to increase the capacity over smaller, busy areas and to cover specific areas that the macrocells could hardly reach due to terrain irregularities. Microcells are also used to provide additional cover for high-demand points in the cities. Recently, interest has arisen in the management of spectrum in a macrocell/microcell cellular system [1–3] and its combination with the selection of optimum MAS for use in each layer with low- and high-mobility users [4].

The problem of obtaining optimum spectrum partitioning between the two layers has not yet been solved and is tightly tied to the user classification strategy. If a classification is available, optimum spectrum partitioning can be achieved, leading to an increase in the system efficiency. Any spectrum partitioning method has to consider the user speed distribution and the functionality of each layer [3]. With regard to user speed, it is important to have a known function describing the velocity of users in both microcell and macrocell because this helps to determine roughly the number of handovers experienced in each layer. It can also tell the system operator whether or not the system will operate fast enough to do all the signaling required to perform a handover for a fast moving user in a microcell.

The functionality of each layer also plays an important role in dividing the spectrum. The task that microcells and the macrocells are required to carry out must be specified to let them carry more or less traffic according to their functions. For example, if the primal role of the macrocell layer is to work as an overflow server only, then this will imply sending most of the traffic to the microcells and the calls that cannot be attended by them will be sent to the macrocells. Therefore, more channels should be given to the microcell layer. Alternatively, if the role of the macrocells is to increase the trunking efficiency of its resources, then the more traffic offered to them the better. A similar reasoning can be given for the case of the functionality of microcells, and their role should be combined with the macrocells to produce mutually advantageous trade-offs.

To summarize, dealing with spectrum partitioning in a hierarchical system needs, firstly, an optimum method of classifying users as fast- or as slow-moving users based on previous users' cell dwell time [3–8]. With an optimum classification strategy, the number of handover attempts will be minimized and the signaling load in the network will therefore be decreased, provided the strategy itself does not imply adding extra load to the network.

In this chapter spectrum partitioning in a simplified two-tier macrocell/microcell cellular system is studied. Optimum spectrum partitioning is aimed at minimizing the blocking probability in each layer, taking into account two distinct user behavior types: slow- and fast-moving users (pedestrians and cars, respectively). The number of slow-moving users will depend on the actual traffic offered to the system and on the speed threshold set for mobiles in the microcells. The teletraffic analysis carried out here is a particular case of the analysis presented in [9].

5.3 Mathematical Approach Used

This study of the performance of a multilayered network assumes that user classification based on its speed is perfect. Hence, after a user has decided or has been informed of to which group it belongs, it will originate a call to the proper layer [3]. The decision on whether a mobile will be classified as a slow- or as a fast-moving user depends on the speed threshold for users in microcells V_0, which in turn is related to the dwell time threshold τ_0 and the cell radius R via [3]

$$\tau_0 = \frac{R\pi}{2V_0} \tag{5.1}$$

Users with speed lower than or equal to V_0 are considered slow-moving users and are attended by microcells in the first place, whereas users with speeds greater than V_0 are considered as fast-moving users and are attended by macrocells only.

Let us assume that the velocity probability density function (pdf) of mobiles in the system $f(v)$ is a triangular function as in [3] and is given by

$$f(v) = \begin{cases} \dfrac{v}{100} & 0 \le v \le 10 \\[4mm] -\dfrac{v}{100} + \dfrac{20}{100} & 10 < v \le 20 \end{cases} \tag{5.2}$$

Then, from $f(v)$, the mean velocities of mobiles in microcells and macrocells are, respectively,

$$E_\mu[v] = \begin{cases} \dfrac{2}{3}V_0 & \text{for} \quad V_0 \le 10 \\[4mm] \dfrac{2}{3}\dfrac{V_0^{\,3} - 30V_0^{\,2} + 1000}{V_0^{\,2} - 40V_0 + 200} & \text{for} \quad 10 < V_0 \le V_L \end{cases} \tag{5.3}$$

$$E_M[v] = \begin{cases} \dfrac{2}{3}\dfrac{V_0^{\,3} - 3000}{V_0^{\,2} - 200} & \text{for} \quad V_0 \le 10 \\[4mm] \dfrac{2}{3}\dfrac{V_0^{\,3} - 30V_0^{\,2} + 4000}{V_0^{\,2} - 40V_0 + 400} & \text{for} \quad 10 < V_0 \le 20 \end{cases} \tag{5.4}$$

In these equations, V_L stands for a upper speed limit set in microcells. This limit is set up because mobiles with speed greater than V_L will not spend sufficient time in the microcell to complete call set-up or handover functions. It could also be a function of the maximum number of handovers per unit of time that a microcell would be able to accommodate.

To proceed with the calculation of the blocking probability in each layer, we study an area covered by a set of W microcells perfectly overlaid by only one macrocell as in [9]. The velocity distribution keeps constant, even when the traffic load in the system increases. Whenever a new call arrives at a microcell, this cell will try to accommodate the call in this layer; but if there is no channel available, the call will be overflowed to the macrocell level. The overlaying macrocell will try to give service to this call, but if there is no resource available the call will be blocked. Once a call has been successfully sent to the macrocell layer, this call will remain in that layer until the end of the communication. A similar procedure is followed by the handover calls. The scenario for call management is shown in Figure 5.1.

In the homogeneous case, mobile stations (MSs) are uniformly distributed over the whole area and all cells in the same hierarchical level are statistically identical. In this way analysis can focus on one cell of each layer only and treat it independently as representative of all cells in the level [9]. FCA is used for both microcells and macrocells for simplicity. Reserved channels are implemented in both the microcells and the macrocells to reduce the handover failure probability. No queuing of calls is therefore considered in the teletraffic analysis. This also excludes the case of "virtual queuing," which occurs in some existing digital systems (GSM, in particular), where a call is not dropped merely because a handover failure has occurred due to the lack of resources in the target cell. These calls are maintained as long as possible, and a new handover is

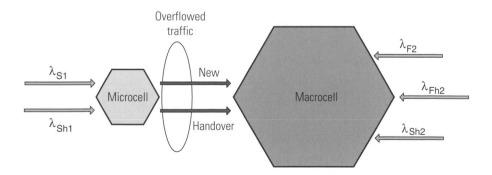

Figure 5.1 Management of calls in the proposed system.

attempted until the signal level reaches a point where successful communication can be kept up (provided the user is still under an overlapping area and the criteria for handover still exists). Excluding queues simplify the analytical method enormously.

New call arrivals into an area (a set of only one macrocell and its overlaid microcells) follow a Poisson process and are divided between two strings depending on the speed threshold: new calls from slow-moving users directed to microcells and new calls from fast-moving users directed to macrocells.

If we denote the total offered traffic load to an area as λ_T, then this could be expressed in terms of the offered traffic to a microcell λ_μ and to a macrocell λ_M as

$$\lambda_T = W\lambda_\mu + \lambda_M \tag{5.5}$$

where

$$\lambda_\mu = \frac{\lambda_T}{W}\left[P(v \leq V_0)\right] \tag{5.6}$$

and

$$\lambda_M = \lambda_T\left[P(v > V_0)\right] \tag{5.7}$$

The service law of communication follows a negative exponential distribution with mean $1/\mu$. The dwell time also follows a negative exponential distribution and is different for each type of user (slow- and fast-moving users) in microcells and in macrocells. This requires the following specific parameters:

- $\lambda_\mu = \lambda_{S1}$ and $\lambda_m = \lambda_{F2}$ are the new call arrival rate to microcells from slow-moving users and the new call arrival rate to macrocells from fast-moving users, respectively.
- λ_{Sh1} and λ_{Sh2} are the handover arrival rate from slow-moving users in a microcell and a macrocell, respectively.
- λ_{Fh2} is the handover arrival rate from fast-moving users in a macrocell.
- $1/\eta_{S1}$ and $1/\eta_{S2}$ represent the mean dwell time for slow-moving users in microcells and macrocells, respectively.
- $1/\eta_{F2}$ is the mean dwell time for fast-moving users in macrocells.

The nomenclature is fairly obvious. For example, λ_{Sh1} is the handover arrival rate made by slow-moving users into a cell in layer 1 (the microcell layer).

For the analysis, circular microcells with radii $R_1 = 300$m are used. Based on this knowledge, the dwell times can be expressed as

$$\frac{1}{\eta_{S1}} = \frac{R_1 \pi}{2 E_\mu [v]} \tag{5.8}$$

$$\frac{1}{\eta_{S2}} = \frac{R_1 \pi \sqrt{W}}{2 E_\mu [v]} \tag{5.9}$$

$$\frac{1}{\eta_{F2}} = \frac{R_1 \pi \sqrt{W}}{2 \mathrm{E_M} [v]} \tag{5.10}$$

5.3.1 Teletraffic Analysis of the Microcell Level

The overall aim is to evaluate the performance of a two-layer HCS with different partitioning and select the one that performs better in terms of maximum carrying of offered load. Two different strategies for handover will be considered. They are the NPS and the RCS, as introduced in Chapter 2. Deriving the model to describe a system using the more generic of these strategies (i.e., RCS) will allow us to model the other strategy (i.e., NPS) without any problems. For these reasons, the formulas for modeling a HCS with two layers using RCS are derived next.

Each microcell has S_1 channels and $S_1 - n_1$ channels are reserved for handover calls only. Considering the state transition diagram of Figure 5.2, we can obtain the blocking probability of each microcell in the usual way [10]

$$P_j = \frac{\left(\dfrac{\lambda_{S1} + \lambda_{Sh1}}{\mu + \eta_{S1}} \right)^j}{j!} P_0 \qquad 0 < j \leq n_1 \tag{5.11}$$

and

$$P_j = \frac{\left(\lambda_{Sh1} \right)^{j - n_1} \left(\lambda_{S1} + \lambda_{Sh1} \right)^{n_1}}{\left(\mu + \eta_{S1} \right)^j j!} P_0 \qquad n_1 < j \leq S_1 \tag{5.12}$$

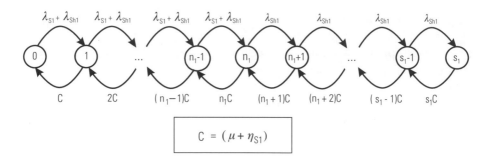

Figure 5.2 State transition diagram for the RCS.

Applying the normalization equation

$$P_0 = \left[\sum_{j=0}^{n_1} \frac{\left(\frac{\lambda_{S1} + \lambda_{Sh1}}{\mu + \eta_{S1}} \right)^j}{j!} + \sum_{j=n_1+1}^{S_1} \frac{\left(\lambda_{Sh1} \right)^{j-n_1} \left(\lambda_{S1} + \lambda_{Sh1} \right)^n}{j!(\mu + \eta_{S1})^j} \right]^{-1}$$ (5.13)

where j is the number of slow-moving users in the microcell.

The probability Pb_1 that a new call from a slow-moving user is blocked from the microcell and thus overflows to the overlaying macrocell is

$$Pb_1 = \sum_{j=n_1}^{S_1} P_j$$ (5.14)

and the probability that a handover call from the same type of user Ph_1 overflows to the overlaying macrocell is

$$Ph_1 = P_{S_1}$$ (5.15)

An iterative approach for determining the handover arrival rate to each microcell is used in much the same manner as in Section 3.3.5. Once the set of probabilities Pb_1 and Ph_1 have been calculated, the mean and variance of the overflow traffic can also be calculated by introducing a fictitious, infinite server overflow group as in [9]. Having M channels in it, M is sufficiently large compared with the load submitted to a microcell. This system could be described by a state (j, N) with equilibrium probability $Q(j, N)$, where j is as defined previously and N is the number of calls present in the overflow group. This system is shown in Figure 5.3. The state equations, once again, can be formulated in the usual way [10].

The equilibrium state probabilities for $0 \leq j < n_1$ and $0 \leq N \leq M$ are

$$[\lambda_{S1} + \lambda_{Sh1} + j(\mu + \eta_{S1}) + N\mu]Q(j, N) \qquad (5.16)$$
$$= (\lambda_{S1} + \lambda_{Sh1})Q(j - 1, N) + (j + 1)(\mu + \eta_{S1})Q(j + 1, N)$$
$$+ (N + 1)\mu Q(j, N + 1)$$

When $n_1 \leq j < S_1$, new call arrivals will overflow to the macrocell layer while handover calls are still accommodated by microcells. Define

$$A_1 = \begin{cases} 0 & if \quad N = M \\ 1 & otherwise \end{cases} \qquad (5.17)$$

Then for $j = n_1$ and $0 \leq N \leq M$

$$[\lambda_{S1}A_1 + \lambda_{Sh1} + j(\mu + \eta_{S1}) + N\mu]Q(j, N) \qquad (5.18)$$
$$= (\lambda_{S1} + \lambda_{Sh1})Q(j - 1, N) + \lambda_{Sh1}Q(j, N - 1)$$
$$+ [(j + 1)(\mu + \eta_{S1})]Q(j + 1, N) + (N + 1)\mu Q(j, N + 1)$$

for $n_1 < j < S_1$ and $0 \leq N \leq M$

$$[\lambda_{S1}A_1 + \lambda_{Sh1} + j(\mu + \eta_{S1}) + N\mu]Q(j, N) \qquad (5.19)$$
$$= \lambda_{Sh1}Q(j - 1, N) + \lambda_{S1}Q(j, N - 1)$$
$$+ [(j + 1)(\mu + \eta_{S1})]Q(j + 1, N) + (N + 1)\mu Q(j, N + 1)$$

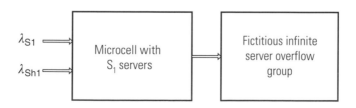

Figure 5.3 Management of calls overflowed from the microcell layer.

and finally, the equilibrium equations for the boundary states $j = S_1$ and $0 \leq N \leq M$

$$[(\lambda_{S1} + \lambda_{Sh1})A_1 + j(\mu + \eta_{S1}) + N\mu]Q(j,N)$$
$$= (\lambda_{S1} + \lambda_{Sh1})Q(j,N-1) + \lambda_{Sh1}Q(j-1,N)$$
$$+ (N+1)\mu Q(j,N+1)$$
$$(5.20)$$

with the following constraint for (18)–(20)

$$Q(j,-1) = Q(j,M+1) = 0 \tag{5.21}$$

and subject to the normalization equation to give a total probability of one

$$\sum_{j=0}^{S_1}\sum_{N=0}^{M} Q(j,N) = 1 \tag{5.22}$$

The mean α_i and the variance v_i of the overflow traffic distribution resulting from an overlaid subordinated cell only can be computed from the state probabilities as [9]

$$\alpha_i = \sum_{j=0}^{S_1}\sum_{N=0}^{M} NQ(j,N) \tag{5.23}$$

$$v_i = \left\{\sum_{j=0}^{S_1}\sum_{N=0}^{M} N^2 Q(j,N)\right\} - \alpha_i^2 \tag{5.24}$$

Then, remembering that each macrocell overlays exactly W microcells, the mean and variance of the total overflow traffic could be found by

$$\alpha_T = \sum_{i=1}^{W} \alpha_i \tag{5.25}$$

$$v_T = \sum_{i=1}^{W} v_i \tag{5.26}$$

5.3.2 Modeling the Overflow Traffic

The total overflow traffic is modeled by an interrupted Poison process (IPP). The way to determine the IPP parameters $(\lambda, \gamma, \omega)$ can be found in [11] and, with revised formulas and a justification, in [9]. λ represents the intensity of the modulated Poisson process, $1/\gamma$ is the mean on-time, and $1/\omega$ is the mean off-time of the random switch that modulates the interrupted Poisson process. The on-time and the off-time random variables follow a negative exponential probability density distribution.

5.3.3 Teletraffic Analysis of the Macrocell Level

The mean overflow traffic arrival rates from the microcells for new calls and handover calls of slow-moving users are given in terms of IPP parameters [11] as

$$L_{S1} \approx \frac{(\gamma + \omega)}{\omega}\left(\lambda_{S1}Pb_1 W\right) \tag{5.27}$$

$$L_{Sh1} \approx \frac{(\gamma + \omega)}{\omega}\left(\lambda_{Sh1}Pb_1 W\right) \tag{5.28}$$

An overlaying macrocell of Figure 5.4 is described by the state (j, k, Z) with equilibrium state probability $P_2(j, k, Z)$, in which j and k are the number of slow- and fast-moving users in the observed macrocell, respectively. Z is the state of IPP random switch taking the value of 1 if the process is on or 0 if the process is off.

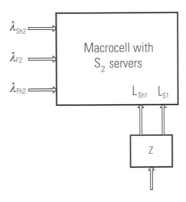

Figure 5.4 Analysis of an overlaying macrocell.

In each macrocell there are S_2 channels, from which $S_2 - n_2$ are reserved for handover calls only. Fast-moving users are also present in the macrocells. They are denoted by k in the following equations, as stated previously.

Consider

$$A_2 = \begin{cases} 0 & if \quad (j+k) = n_2 \\ 1 & otherwise \end{cases} \tag{5.29}$$

$$A_{h2} = \begin{cases} 0 & if \quad (j+k) = S_2 \\ 1 & otherwise \end{cases} \tag{5.30}$$

Then for: $0 \le (j+k) \le n_2$

$$\left[\lambda_{Sh2} + \lambda_{F2}A_2 + \lambda_{Fh2} + j(\mu + \eta_{S2}) + k(\mu + \eta_{F2}) + \omega \right] P_2(j,k,0) \tag{5.31}$$
$$= \gamma P_2(j,k,1) + \lambda_{Sh2} P_2(j-1,k,0) + (\lambda_{F2} + \lambda_{Fh2}) P_2(j,k-1,0)$$
$$+ \left[(j+1)(\mu + \eta_{S2}) \right] P_2(j+1,k,0) + \left[(k+1)(\mu + \eta_{F2}) \right] P_2(j,k+1,0)$$

and with the constraint for the switch off

$$P_2(-1,k,0) = P_2(j,-1,0) = 0 \tag{5.32}$$

$$\tag{5.33}$$

$$\left[(L_{S1} + \lambda_{F2}) A_2 + L_{Sh1} + \lambda_{Fh2} + \lambda_{Sh2} + j(\mu + \eta_{S2}) + k(\mu + \eta_{F2}) + \gamma \right] P_2(j,k,1)$$
$$= \omega P_2(j,k,0) + (\lambda_{Sh2} + L_{S1} + L_{Sh1}) P_2(j-1,k,1)$$
$$+ (\lambda_{F2} + \lambda_{Fh2}) P_2(j,k-1,1)$$
$$+ \left[(j+1)(\mu + \eta_{S2}) \right] P_2(j+1,k,1) + \left[(k+1)(\mu + \eta_{F2}) \right] P_2(j,k+1,1)$$

with the constraint for the switch on

$$P_2(-1,k,1) = P_2(j,-1,1) = 0 \tag{5.34}$$

For $n_2 < (j+k) \le S_2$

$$\left[\left(\lambda_{Sh2} + \lambda_{Fh2}\right) A_{h2} + j\left(\mu + \eta_{S2}\right) + k\left(\mu + \eta_{F2}\right) + \omega\right] P_2\left(j, k, 0\right) \qquad (5.35)$$
$$= \gamma P_2\left(j, k, 1\right) + \lambda_{Sh2} P_2\left(j - 1, k, 0\right) + \lambda_{Fh2} P_2\left(j, k - 1, 0\right)$$
$$+ \left[\left(j + 1\right)\left(\mu + \eta_{S2}\right)\right] P_2\left(j + 1, k, 0\right) + \left[\left(k + 1\right)\left(\mu + \eta_{F2}\right)\right] P_2\left(j, k + 1, 0\right)$$

with the constraint $P_2(j, k, Z) = 0$ if $(j + k) > S_2$, with $t = 0$ or $t = 1$

$$\left[\left(\lambda_{Sh2} + \lambda_{Fh2} + L_{Sh1}\right) A_{h2} + j\left(\mu + \eta_{S2}\right) + k\left(\mu + \eta_{F2}\right) + \gamma\right] P_2\left(j, k, 1\right) \qquad (5.36)$$
$$= \omega P_2\left(j, k, 0\right) + \left(\lambda_{Sh2} + L_{Sh1}\right) P_2\left(j - 1, k, 1\right) + \lambda_{Fh2} P_2\left(j, k - 1, 1\right)$$
$$+ \left[\left(j + 1\right)\left(\mu + \eta_{S2}\right)\right] P_2\left(j + 1, k, 1\right) + \left[\left(k + 1\right)\left(\mu + \eta_{F2}\right)\right] P_2\left(j, k + 1, 1\right)$$

with $P_2(j, k, Z) = 0$ if $(j + k) > S_2$ with $t = 0$ or $t = 1$.

The normalization equations for this case are given by

$$\sum_{j=0}^{S_2}\sum_{k=0}^{S_2} P_2\left(j, k, 1\right) = \frac{\omega}{\left(\gamma + \omega\right)} \qquad (5.37)$$

$$\sum_{j=0}^{S_2}\sum_{k=0}^{S_2} P_2\left(j, k, 0\right) = \frac{\gamma}{\left(\gamma + \omega\right)} \qquad (5.38)$$

When the state probabilities are found, an iterative approach is used to assess the handover arrival rates. As is usually done, we start with an estimate of the values of the handover arrival rates and then calculate the handover departure rates. We proceed in this way until the state probabilities satisfy the normalization equations and the handover arrival rates equal the handover departure rates.

The probability that a new call (of any type) is denied access to a macrocell and thus blocked from the hierarchical structure, Pb_2, is

$$Pb_2 = \sum \sum_{(j+k)=n_2}^{S_2} \left[P_2\left(j, k, 0\right) = P_2\left(j, k, 1\right)\right] \qquad (5.39)$$

and the conditional probability that the macrocell will not be able to serve a new call when a overflowed slow-moving user call arrives is

$$PB_2 = \sum \sum_{(j+k)=n_2}^{S_2} P_2\left(j, k, 1\right) \qquad (5.40)$$

The probability that a handover call is denied access to a channel in a macrocell and hence forced to terminate is

$$Ph_2 = \sum \sum_{(j+k)=S_2} \left[P_2(j,k,0) + P_2(j,k,1) \right] \tag{5.41}$$

and the conditional probability that the macrocell will not be able to serve a handover call when a overflowed slow-moving user handover call arrives is

$$PH_2 = \sum \sum_{(j+k)=S_2} P_2(j,k,1) \tag{5.42}$$

The total blocking probability, Pb_{ST}, for a new call origination from a slow-moving user is

$$Pb_{ST} = Pb_1 PB_2 \tag{5.43}$$

Note that the blocking probability for fast-moving users, which travel only in the macrocell layer, will be larger than the blocking probability for slow-moving users, which travel in both layers.

5.4 Example and Discussions

Using the formulas derived previously to calculate the blocking probability in each layer (Pb_1 for microcells and Pb_2 for macrocells), the speed threshold could be found based on the load present. As in [3], the speed threshold V_0 is set such that the grade of service (GOS) of 2% is met both in the macrocell and microcell so a comparison of results is feasible when overflow traffic is taking into account when it is not. However, V_0 is set such that the traffic partitioned into the macrocell and the microcell makes Pb_2 = GOS and Pb_1 < GOS. V_0 can be increased up to the point that either Pb_1 = GOS or $V_0 = V_L$. To ensure that the number of handovers processed in the microcell level is minimized, the minimum possible value for V_0 is selected such that the constraints explained previously are fulfilled.

In the following numerical example, 10 channels are initially assigned to each microcell and 10 channels to each macrocell. The upper speed limit is set to 17 m/s and $1/\mu$ = 2 min. For simplification purposes, the reuse pattern is the same for macrocell and microcell levels and is equal to 7. In this situation, an increase in the number of channels in the microcell level will represent a reduction of exactly the same number of channels in the macrocell level. The spectrum partitioning technique to follow is that described in [3].

5.4.1 Optimum Spectrum Partitioning Using NPS

Let us consider first the case when no reserved channel is kept in the microcell level or in the macrocell. Calculation starts with the assumption of equal partitioning of the spectrum. It is (10, 10), with 10 channels available in each level of the hierarchy. This combination is limited by the GOS constraints rather than the upper speed limit: at V_0 = 16.95 m/s, $Pb_1 = Pb_2$ = GOS. The exploration continues by increasing the number of channels in the microcell level by 1, which produces the partitioning (9, 11). This combination is limited by the upper speed limit and is able to deal with 41.25 Erl offered to an area. Calculation is continued in the same way until optimum partitioning is reached when seven channels are given to the macrocell and 13 to the microcell level. This is also limited by the upper speed limit but can handle a load of 44 Erl in the area, meeting the GOS constraints. The pair (6, 14) and (5, 15) can handle 43 Erl and 37 Erl, respectively.

This optimal partitioning differs from that found in [3] because of the inclusion of overflow traffic from the microcell to the macrocell level. Figure 5.5 depicts the results of the procedure outlined in this section. The

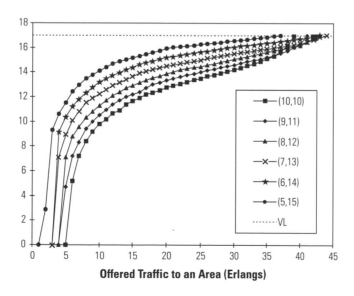

Figure 5.5 System traffic capacity with several possible spectrum partitioning combinations using the NPS in both layers. ©1999 IEEE. Reprinted, with permission, from [12].

greater the number of channels we give to the microcell level, the larger the capacity it can handle (if there was no upper speed restriction). But, since there was a limit, an intermediate point had to be found.

5.4.2 Optimum Spectrum Partitioning Using RCS

When the RCS is used in both layers, the results are quite different from those previously stated. Figure 5.6 shows the evaluation for the possible ways of spectrum partitioning when there is one reserved channel in each level. In this case, calculation also starts with the initial partitioning (10, 10) and continues until we find that no advantages are gained by changing the partition ratio. With the RCS, as the number of channels for new calls and overflow traffic in the macrocell level is decreased, the GOS set is met at low values of offered traffic. A system like this deals with lower loads than a system with NPS. Obviously, this is not desirable in any real system.

Because the microcell layer has reserved channels for handover calls, the handover failure rate probability in this level is decreased for slow-moving users but creates an increase in the number of new calls from this type of user that are overflowed to the macrocell level. This, in turn, increases the blocking probability in the macrocell layer.

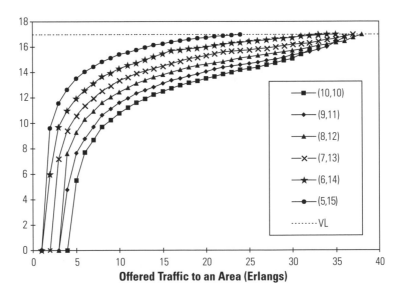

Offered Traffic to an Area (Erlangs)

Figure 5.6 System traffic capacity with several possible spectrum partitioning combinations using the RCS in both layers.

This system is limited by the GOS constraints for the cases (10, 10) and (9, 11). The optimum partitioning is (8, 12).

5.4.3 RCS and NPS—Optimum Spectrum Partitioning

Figure 5.7 shows the results of the search for the spectrum partitioning when the RCS is used in the microcell level (only one reserved channel) and NPS is used in the macrocell level. For this case, the optimum partitioning is also (7,13). No significant differences were found in the traffic-carrying capacity in comparison with the NPS scheme in both layers (about 1.7 Erl/area). This means that the system can decrease the handover failure rate for slow-moving users without much of a decrease in the maximum carried load in the area. The explanation for this phenomena is that the overflow traffic coming from the lower layer saturates the upper layer in much the same manner as in the previous case, when only one channel was reserved for handover calls.

Figure 5.8 depicts the results for the opposite case: NPS in the microcell level and RCS in the macrocell level. As in the case of RCS in both layers, the optimum partitioning is (8,12), with an increase in the overall traffic-carrying capacity in the area (1.5 Erl) and a reduction in the number of handover failure rate for fast-moving users (mainly) in the macrocell level.

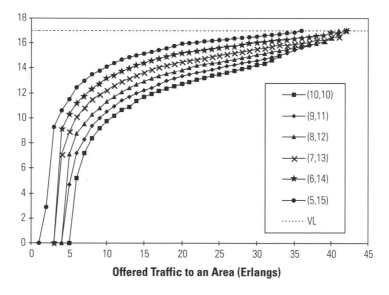

Figure 5.7 System traffic capacity with several possible spectrum partitioning combinations using NPS in the macrocells and RCS in the microcells.

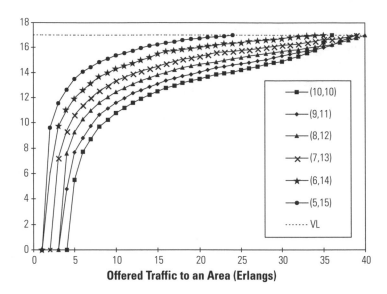

Figure 5.8 System traffic capacity with several possible spectrum partitioning combinations using RCS in the macrocells and NPS in the microcells.

There is a trade-off between the trunking efficiency of channels in macrocells and their role as an overflow group of servers since channels in macrocells should be available for overflowed calls but, at the same time, should be occupied most of the time in order to increase channel efficiency. This trade-off is tightly tied to the handover failure rate at the macrocell level because the more time channels are busy attending new and overflow calls, the higher the blocking rate for handovers will be.

The way slow- and fast-moving users are treated in this spectrum partitioning technique creates some problems. For instance, consider the following case. Assume that a slow-moving user does not find a free channel upon its arrival to a microcell. This user will be overflowed to the macrocell layer to try to find a channel available. If it succeeds, then it will remain in the macrocell layer until the end of its communication provided it has successful handovers when crossing macrocells borders. If under these conditions a fast-moving user arrives at the cell where the slow-moving user was overflowed and there are no channels available, the call attempt made by the fast-moving user will be blocked. This case is worsened when there are channels available in the microcells coverage area the slow-moving user is traversing, because a channel reallocation could have been performed to transfer the call of this slow moving user down to the microcell it is traversing. This last process is called hand-down.

If a hand-down strategy is considered in the spectrum partitioning performed in this chapter, the results would change and the teletraffic analysis would get more complicated. Following a similar reasoning, if an HCS allowed fast-moving users to initiate calls in the lower layers, great improvements could be expected if the opposite technique (hand-up) is used. Having understood what a hand-down is, it is easy to understand what a hand-up is and why it would improve the system performance.

5.5 Summary

A study has been made of spectrum (channel) partitioning in a hierarchical system with two layers only, a technique that could be very popular in future personal communications services networks. Overflow traffic has been considered in the calculations, and perfect layer selection based on user speed has been supposed. Due to these considerations, results for optimum spectrum partitioning are different from previous studies. This could be translated into important improvements in system capacity and a reduction in signaling load.

This study has also taken into account the trade-off between the trunking efficiency of the channels in the macrocell and their role as an overflow group of channels. However, this trade-off cannot be completely solved until the possibilities of reuse of channels between layers and of moving channels among cells and between layers (complete DCA) are available.

References

[1] Riva, G., "Preliminary Results on Traffic Overflow from Microcells to Macrocell," *IEEE Electronics Letters*, Vol. 28, No. 15, July 16, 1992, pp. 1462–1463.

[2] Frullone, M., G. Riva, P. Grazioso, and C. Carciofi, "Analysis of Optimum Resource Management Strategies in Layered Cellular Structures," in *Proc. Int. Conf. on Universal and Personal Commun., ICUPC'94*, pp. 371–375, San Diego, CA, USA. 1994.

[3] Yeung, Kwan L., and Sanjiv Nanda, "Channel Management in Microcell/Macrocell Cellular Radio Systems," *IEEE Trans. Vehicular Tech.*, Vol. 45, No. 4, Nov. 1996, pp. 601–612.

[4] Chih-Lin, Larry J. Greenstein, and Richard D. Gitlin, "A Microcell/Macrocell Cellular Architecture for Low and High Mobility Wireless Users," *IEEE J. Select. Areas Commun.*, Vol. 11, No. 6, Aug. 1993, pp. 601–612.

[5] Lagrange, X., and P. Godlewski, "Performance of a Hierarchical Cellular Network with Mobility Dependent Hand-Over Strategies," in *Proc. IEEE Vehicular Technol. Conf. VTC'96*, pp. 1868–1872, Atlanta, GA, USA, April 28–May 1, 1996.

[6]   Wan Sung, Chi, and Wing Shing Wong, "User Speed Estimation and Dynamic Channel Allocation in a Hierarchical Cellular System," in *Proc. IEEE Vehicular Technol. Conf. VTC'94*, pp. 91–95, Stockholm, Sweden, June 1994.

[7]   Benveniste, Mathilde, "Cell Selection in Two-Tier Microcellular/Macrocellular Systems," in *Proc. IEEE GLOBECOM'95*, pp. 1532–1536, Singapore, Nov. 13–17, 1995.

[8]   Shum, Kenneth W., and Chi Wan Sung, "Fuzzy Layer Selection Method in Hierarchical Cellular Systems," in *Proc. IEEE GLOBECOM'96*, pp. 1049–1053, London, UK, Nov. 18–22, 1996.

[9]   Hu, Lon-Rong, and Stephen Rappaport, "Personal Communication Systems Using Multiple Hierarchical Cellular Overlays," *IEEE J. Select. Areas Commun.*, Vol. 13, No. 2, Feb. 1995, pp. 406–415.

[10] Cooper, R. B., *Introduction to Queueing Theory*, 3rd edition, Washington D.C.: CeePress, 1990.

[11] Kuczura, A., "The Interrupted Poisson Process as an Overflow Process," *Bell System Technical J.*, Vol. 52, Mar. 1973, pp. 437–448.

[12] Ortigoza-Guerrero, Lauro, and A.H. Aghvami, "On the Optimum Spectrum Partitioning in a Microcell/Macrocell Cellular Layout with Overflow," *Procedings of the IEEE GLOBECOM '97*, pp. 991-995, Phoenix, AZ, November, 1997.

6

Channel Assignment in Cellular Systems with HCSs

This chapter looks at the characteristics and use of a DDCA strategy.

In the previous chapter, spectral partitioning for a HCS formed of two layers of cells was analyzed and simulated. The partitioning was done in terms of the pdf of the speed of users and in terms of the expected blocking probability in each layer. This meant that each layer was also given a fixed bandwidth. However, this spectrum partitioning technique has no flexibility for readjusting the actual partitioning or for sharing resources between layers when one of the parameters used to obtain the partitioning suddenly changes (e.g., velocity of users).

The spectrum partitioning technique described in Chapter 5 creates an initial and long-term division of resources between layers and is used as such in cellular systems. So, given a pdf describing the speed of the users, the spectrum partitioning is performed once every day, week, and month, for example, and is updated by the system planning engineers as often as is needed or as often as possible. Hence, this technique cannot adapt itself to sudden changes in the pdf of speed of users and hot spots, for example.

To cope with sudden changes of traffic load in each layer, after spectrum partitioning is been performed initially, a way of sharing resources is needed that continues to do this automatically, not only between cells of the same layer but also between cells of different layers of the hierarchy. Due to their unique characteristics, DCA strategies provide the flexibility to achieve this, as explained in Chapter 2, and are one of the best solutions for coping with sudden changes in traffic.

In this chapter, a method is explained that takes changes in the traffic distribution in and between layers into account using a DDCA strategy. The aim of this CAS is to provide the spectrum partitioning strategy, proposed in the previous chapter, with the flexibility to share resources not only between cells of the same layer but between cells of different layers.

Since the main objective of this chapter is to show the reader the behavior of the DDCA itself and to what extent a DDCA can solve the aforementioned problem, a very simple multilayer system is used in its evaluation. However, ways to extrapolate the results obtained to more complex systems can also be found later in the chapter.

As in the previous chapter, the HCS in which the DDCA is to be evaluated by means of computer simulations, consists of a microcell layer overlaid by one macrocell layer. As before, users are identified as fast- and slow-moving users according to their speeds and are directed to an optimum layer: fast-moving users are served by the macrocell layer, whereas the slow-moving users are served by the microcell layer. Again, overflow traffic is considered but so are hand-downs (i.e., a handover from the macrocell to the microcell layer). The DDCA proposed is compared with two other strategies: simple FCA and an enhanced version of FCA with overflow traffic, hand-downs, and time slot reallocations.

6.1 DCA in HCS

In HCS macrocells are used to provide large area coverage with a low-cost infrastructure, meanwhile microcells are used to cope with traffic hot spots to provide sufficient capacity in the network. In HCS, the need to distinguish fast from slow users is a main concern, as explained in the previous chapter. Users' classification in HCS commonly creates two groups: slow-moving users and fast-moving users. The known methods for classifying users are based on residual cell dwell time [1], power level offset [2], exponential averaging [3], and fuzzy layer selection [4]. These are described as follows:

- In the *residual cell dwell time* strategy [1] all calls are initially attended by microcells. Assuming that the call is not completed and requires a handover, it goes either to a microcell or macrocell depending on the amount of time the call stays in the originating cell until the handover is required. If the time the user spent on the originating microcell is longer than a predefined threshold, the call will be handed over to a neighboring microcell, otherwise, it will be sent to a macrocell [3].

- In the *power level offset* strategy [2], when a mobile enters a microcell, a negative offset on the power level received from the microcell's BS is set for a time threshold. If a mobile makes a call origination before the expiration of this time threshold the user originates to the overlaying macrocell, otherwise it will originate to the microcell it is traversing.

- In the *exponential averaging* strategy [3] it is proposed that the mobile must track its own mobility by collecting its past microcell sojourn times. Macro/microcell selection is based on the estimate of the (local) mean sojourn time that is obtained from collected values of the past sojourn times. If these are above a certain threshold the user is allocated to its microcell; if not then it goes to the macrocell.

- Finally, in the *fuzzy layer* selection [4] the problem of classifying users based on their speeds is tackled using past cell dwell time and occupancies in the target micro- and macrocells. This strategy also tries to select the target cell with the lower occupancy in order to prevent call blocking. These heuristics are incorporated in a fuzzy layer selection algorithm, which is devised in terms of fuzzy set and fuzzy logic.

The aim of all these strategies is to decrease the large number of handovers that would be carried out by fast-moving users in microcells and the number of handovers between layers due to an initially incorrect layer selection. They also aim to decrease the number of times a slow-moving user is given a channel in a macrocell due to erroneous classification.

As explained in Section 5.3.3 this last error decreases the spectrum efficiency considerably and increases the fast-moving users' dropping and blocking probabilities when no hand-down is available, especially when the main role of the macrocells is to work as overflow groups because macrocell channels are precious resources needed to fulfill the role they were assigned for them, that is, attend overflow traffic and fast moving users. When a variation to the traffic offered to this layer is produced (due to a users' classification error) the number of channels available for fast-moving users is decreased; this is reflected by an increase in the call blocking they experience.

In a HCS consisting of microcells and macrocells, the most common way of providing resources to the layers is to divide the available spectrum between the two layers of cells (frequency splitting): one set of frequencies for the microcell layer and one set for the macrocell level. However, this frequency-splitting technique has the disadvantage that the suppression of the interband interference may not be sufficient to provide adequate SIR at all times [5]. Nevertheless, this technique is most commonly used nowadays and is relatively simple.

One point of interest is the need to increase the spectral efficiency. To meet this point, intralayer and interlayer spectrum sharing are critical as is the use of DCA strategies [6]. It is almost a requirement to implement a DCA in a HCS using a distributed strategy in which the degree of communications amongst BSs is minimum. The solution would be suboptimal in terms of spectrum efficiency, but it would be easy to implement and reduce the signaling load when compared with centralized DCA schemes. Also, in HCSs it is required that the resource allocator be in the BSs rather than in the switches. This makes a distributed DCA the best candidate for spectrum management.

As mobile teletraffic demand increases dramatically, interest in second and third generation cellular system structures with macrocell overlays has emerged [7]. This general interest has extended to any system with FDMA/TDMA or CDMA as the main MAS or any combination of them. In Europe, five different MASs were presented to SMG2 as possible candidates for UTRA but only two of them were selected, as explained in Chapter 4. The proposed MAS were: (1) WB-CDMA, (2) WB-TDMA, (3) WB-TD/CDMA, (4) orthogonal frequency division multiplex access (OFDMA) and (5) ODMA. Neither option (2), nor (4), nor (5) managed to make it to the end of the selection procedure. The TDMA proposal did not offer anything new, and the only advantage it had over their competitors was that GSM was based on it. OFDMA was not selected because there was no real system operating it to prove that OFDMA actually worked and it lacked backup from strong manufacturers.

In this chapter a DDCA for intralayer spectrum sharing in an HCS environment with option(2) (FDMA/TDMA) (with the prospect of being used in a system with option (3) as MAS) is presented. The evaluation of the DCA strategy in a FDMA/TDMA system does not restrict its use in systems using other MASs, however. The technique of spectrum partitioning presented in the previous chapter is used to generate two different bands for each layer, and the DCA is applied at both the carrier level and, to give further improvements in the spectrum efficiency, at the time slot level. Intracell and intercell handovers (intralayer handovers) as well as hand-downs and hand-ups (interlayer handoffs) are considered.

6.2   System Description—Two-Layer HCS with Slow and Fast Users

Figure 6.1 depicts a set of W microcells overlaid by one and only one macrocell. In the geographical area covered by this set of cells, there exist users with different speeds. They are classified as fast- and slow-moving users. Each layer in the

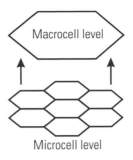

Figure 6.1 The HCS basic scenario.

network is given a subset of carriers from the total available spectrum as a result of applying the spectrum-partitioning technique (as done in the previous chapter). Each microcell and each macrocell is given a set of carriers according to the carrier assignment strategy in use and no channels are reserved for handover.

The way the system deals with incoming calls is described next. This can also be seen as a description of the FCA strategy.

- A new call originated from a slow-moving user will be directed toward its preferred microcell only. If there is no time slot available to serve the call, it will be overflowed to its preferred macrocell (to the next hierarchical level). The preferred cell in each layer is that cell that receives the strongest signal from the mobile in the referred level. If there is still no time slot available, the call will be blocked. A new call originated from a fast-moving user will be served by its preferred macrocell. When there is no resource available to give service to the incoming call in the macrocell level, the call will be blocked.

- Whenever a slow-moving user leaves its preferred microcell domain to enter another microcell, which in turn will become the new preferred microcell, a handover will be performed. The target microcell will try to accommodate the handover call in an idle time slot, otherwise the call will be overflowed to its preferred macrocell. If there is still no time slot available, the call will be dropped. When the handover is performed either by a fast-moving user or by a slow-moving user in the macrocell level, the target macrocell will try to assign an idle time slot. If there is no time slot available, the call will be dropped.

In real systems, operators try to keep ongoing calls at the expense of new calls if necessary and, when a handover fails, the call remains on the original channel/resource/time slot until either a successful handover is accomplished or the signal level at one end becomes unacceptable. This point is usually reached after the threshold for handover begins. However, to simplify the simulation, it is assumed that every time a user reaches the border of a cell it will immediately attempt a handover and, if it does not find a resource available, it will be dropped immediately. This assumption is normally made in analytical methods (teletraffic analysis) and in some computer simulators. Please note that this situation does not happen in real systems.

Users are only tracked at the microcell level. All users are given a cell dwell time for the microcell area in which they are currently traveling. If they are fast-mobile users, at the end of each session (the time that the user spends in a microcell) a check is made to see whether or not the user has changed the macrocell in order to produce a handover to a new macrocell.

6.3 Carrier Allocation Strategies

Four different strategies are considered: classic FCA, FCA with intracell channel reassignments (FCAR), and Hierarchical DCA at the carrier level (HDCA-I) and at the time slot level (HDCA-II). The channel allocation strategies are explained next.

6.3.1 FCAR

Since resources in the macrocell layer are scarce and should be updated according to the traffic load [8], hand-down capability has been added to the classic FCA strategy to increase the possibility of channel acquisitions in the macrocell layer for call origination by fast-moving users. A hand-down is defined as the termination of service given to a particular user in a cell of a layer of the HCS (in this case, the macrocell layer) to continue being served by a cell in a layer of lower hierarchy (in this case the microcell layer), that is, a reallocation of a user from the macrocell to the microcell layer. The hand-down procedure (HDP) is similar to the take back request described in [9] and consists of sending back a slow-moving user served by a macrocell to its preferred microcell to release the resources in the macrocell. In a two-layer HCS, a hand-down might be necessary when, for example, a slow-moving user is being served by a macrocell and resources are needed to attend a new call from a fast-moving user arriving at that macrocell. This is possible because a slow-moving user served by a macrocell continuously monitors the microcell it is traversing.

6.3.1.1 Acquisition Policy for FCAR

The acquisition policy for FCAR follows when a new call or a handover call from a fast-moving user or a handover call from a slow-moving user is generated in a cell at the macrocell level.

- If there is one idle time slot in the nominal carriers, then the idle time slot is assigned to the call (preference will be given to busy carriers with one or more busy time slots).

- If there is no time slot available on nominal carriers but there are calls from slow-moving users served by time slots on carriers that belong to the macrocell, then a hand-down is performed to release resources in the macrocell. The idle time slot is given to the new incoming call, and the HDP is applied as many times as possible afterward to try to release the carrier.

- If no time slot is available at all, the fast-moving user call is blocked from the system.

6.3.1.2 Release Policy for FCAR

The release policy for the FCAR strategy when a call from a slow-moving user is finished at a microcell is described as

- The time slot(s) is(are) put in the idle state; and if there are no more calls attended on the same carrier, then the carrier is released.

- If the overlaying macrocell is attending slow-moving users traversing the microcell, then all the possible hand-downs will be performed to try to release the macrocell carrier.

According to these descriptions, the FCAR strategy has overflow and the hand-down capabilities.

6.3.2 Hierarchical Dynamic Carrier Allocation

In [10] a CAS was proposed that makes use of channel ordering in each cell. This strategy was called BCO. In [11] a modification of this scheme was presented and called SDCA. In this last strategy carrier ordering rather than channel ordering was used. In Chapter 4, the SDCA strategy was used to assess the spectrum efficiency of two different MASs for UTRA. SDCA was applied not at the carrier level but at the code level (a code within a time slot). In this chapter, the proposed HDCA-I complements the SDCA strategy to account for HCSs. Similar to the strategy in [11], an ordered list of carriers rather than channels or resources is used in HDCA-I.

The features that HDCA-I has added to SDCA to make it suitable to operate in HCSs are the carrier allocation policy (Section 6.3.1.1) and the carrier release policy (Section 6.3.1.2) used by FCAR. Also, HDCA-I uses the basic call management methods described in Section 6.2. An important thing to notice here is that, in HDCA-I, the aforementioned policies are not applied to "nominal carriers" but to all the carriers that a cell possesses.

As for the SDCA strategy, the HDCA-I scheme does not need an exchange of information within the interference neighborhood. The busy/idle status of carriers can be determined by passive nonintrusive monitoring at each BS of a particular hierarchy level. In fact, this strategy is a distributed strategy and is suboptimal since every cell has access to partial information only [11]. However, the distributed computation and the reduced communication make it feasible for HCSs, where the signaling load among BSs is required to be minimum.

6.3.3 Hierarchical DCA Applied at Time Slot Level—HDCA-II

Each time a BS captures a carrier due to the DCA policy (SDCA), that carrier and all the associated slots become nominal channels to that cell (only for the period of time it is used in the cell) but, eventually, using only a few and leaving all remaining slots unused and unavailable to others [12]. For this reason, the DCA strategy proposed in the previous section is also applied at the time slot level and called HDCA-II. Undoubtedly, this will require near-perfect synchronization among all the BSs in the network. Nevertheless, according to [12], this topic should not be a problem since it is expected that the dynamic resource allocation functionality as well as the control mechanisms be housed at the BSs in the near future. Putting more intelligence into the BSs is a current trend in mobile networks that should resolve this complexity issue in the very near future.

Figure 6.2 depicts the resource allocation procedure for all the strategies described when a slow-moving user is asking for service in a microcell. The resource allocation phase is, of course, applied at the time slot level for the HDCA-II strategy and at the carrier level for FCA, FCAR, and HDCA-I. When a new call attempt is generated at the macrocell level, either by a fast- or a slow-moving user, the resource allocation policy is depicted by Figure 6.3. The resource release policy is shown in Figure 6.4. It is applicable for all

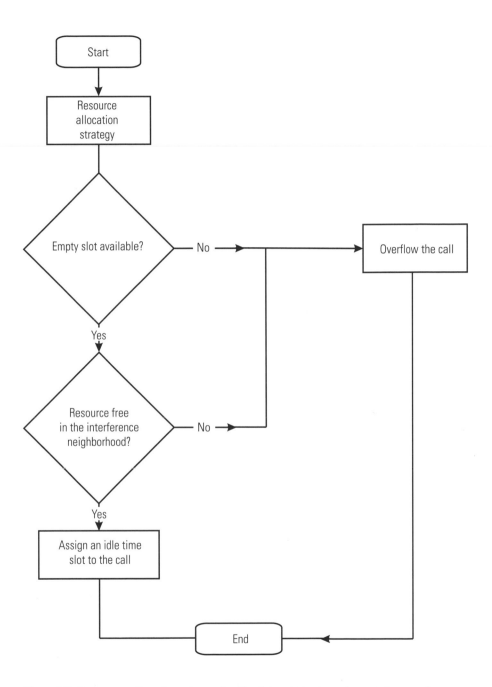

Figure 6.2 Resource allocation phase for the four strategies when a slow-moving user requests service in the microcell level. For HDCA-II the carrier allocation strategy is performed at the time slot level.

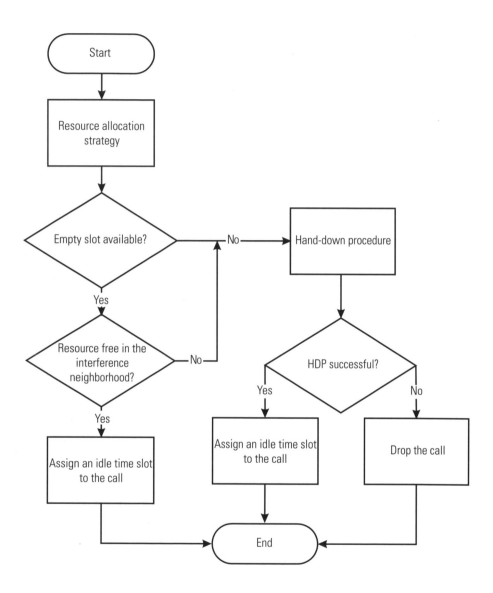

Figure 6.3 Carrier allocation phase when a fast-moving user requests service at the macrocell level (not all the steps are applicable to the four strategies). For HDCA-II the carrier allocation strategy is performed at the time slot level.

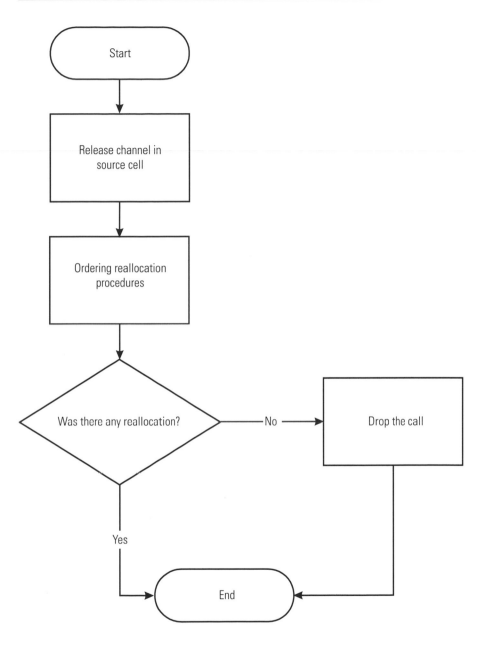

Figure 6.4 Carrier release policy.

the strategies with some exceptions: the HDP is not applicable to FCA and the ordering reallocation procedures are only performed by HDCA-I and HDCA-II. The reallocation procedure has to do with returning borrowed resources to the owner cells.

6.4   Network Scenario Used in Simulations

To evaluate the four strategies described in Section 6.3, a large HCS is simulated by using a wraparound topology. The microcell layer consists of 49 cells (to eliminate the boundary effect that occurs in an unwrapped topology) perfectly overlaid by 7 macrocells. The reuse pattern in simulation is seven for FCA and FCAR. This reuse pattern is used in both the microcell and the macrocell levels. Hexagonal cell geometry is used. The mobility behavior of mobiles in the simulation is described by a two-dimensional random walk as used in [13]. In this model, a mobile stays in the coverage area of a gateway for a period of time that has an exponential distribution with mean $1/\eta$. Then the mobile moves into one of its six neighbors with probability $1/6$. The proposed microcellular simulation environment is that used in Chapter 3 and shown in Figure 3.3. In addition, every cluster of seven cells is overlaid perfectly by a macrocell to give a HCS scenario. So, cells 10 to 16 are overlaid by microcell 1 and so on. Users are tracked in the microcells, hence also determining which macrocell is overlaying them. The traffic pattern shown in Figure 3.4 is used to evaluate the strategies considered in this study under nonuniform traffic load.

6.5.   Examples and Discussion

This section contains results obtained by computer simulations of uniform and real nonuniform voice traffic patterns in the network proposed in Figure 3.3 to verify the methods described earlier in the chapter. Although the optimum spectrum partitioning depends on the current load of the system, the ratio of fast- to slow-moving users, and the selected GOS [8], the actual chosen number of carriers in each microcell is fixed at 12 and the number of carriers in each macrocell at 8 (this was the optimum partitioning found in the previous chapter for the case when NPS is used in both the microcell and the macrocell layers). Every carrier has three time slots.

Only one type of service is considered in the simulations. This service is voice and it is assumed to occupy a single time slot. The new call arrival rate and the handover call arrival rate follow a Poisson process and for all the cases the mean call holding time $1/\mu = 2$ min and the mean dwell time in microcells, $1/\eta$, is 5.83 min and 0.83 min for slow- and fast-moving users, respectively. 75% of new call attempts are made by slow-moving users to microcells and the rest by fast-moving users to macrocells. The blocking probability P_b, the handover failure probability P_h, and the probability of forced termination P_{ft} have been chosen as the parameters for a fair comparison between strategies.

6.5.1 Uniform Traffic

The set of Figures 6.5 to 6.13 compare the performances of FCA, FCAR, HDCA-I, and HDCA-II under symmetric (uniform) traffic load. It can be seen from Figure 6.5 that a substantial reduction in the overall new call blocking, P_{bt}, is achieved by using a DCA strategy rather than FCA. Channel reassignments give a large improvement to the classic FCA, as demonstrated by the FCAR strategy performance, but it is not enough to outperform HDCA-I or HDCA-II. As expected, the possibility of HDCA-II working at the time slot level has made HDCA-II's performance better than that of any other strategies over the entire evaluation range. Due to overflow traffic from slow-moving users, the new call blocking probability P_b at the macrocell level is not equal

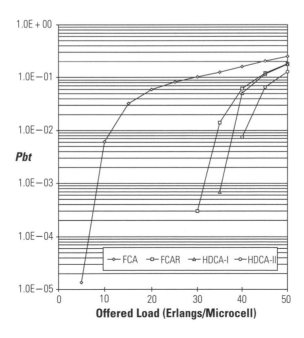

Figure 6.5 Overall blocking probability as a function of the offered load in each microcell.

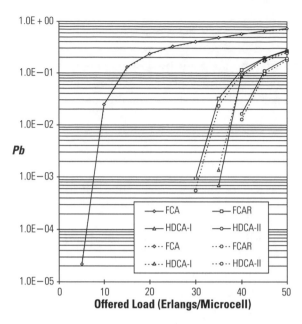

Figure 6.6 Blocking probability in the macrocell layer (bold line) and for fast-moving users (dashed line) in the system versus the offered load in each microcell.

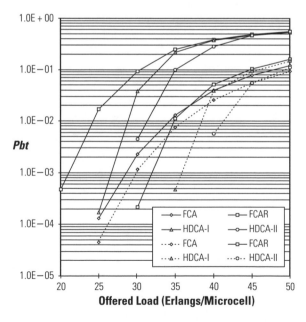

Figure 6.7 Blocking probability in the microcell layer and experienced by slow-moving users in the system (bold and dashed lines respectively) as a function of the offered load in each microcell.

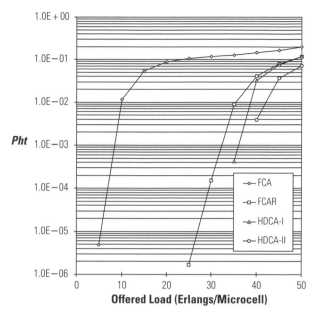

Figure 6.8 Overall probability of handover failure in the system experienced by both types of user as a function of the offered load in each microcell.

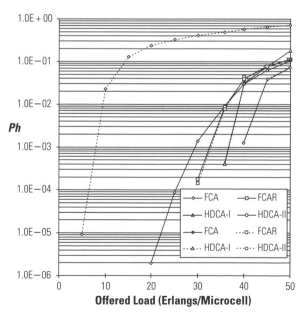

Figure 6.9 Probability of handover failure for slow-moving users and fast-moving users (bold and dashed lines, respectively) versus the offered load in each microcell of the system.

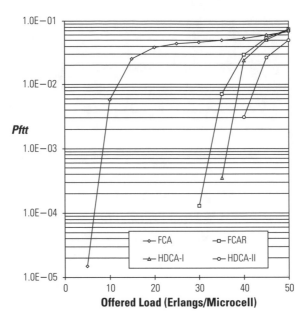

Figure 6.10 Total forced termination probability versus the offered load in each cell of the first hierarchical level.

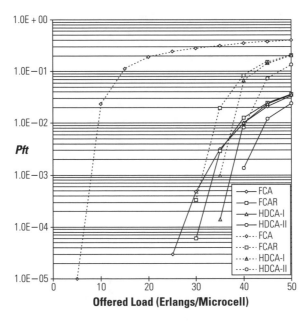

Figure 6.11 Forced termination probability found by for slow-moving users (bold lines) and fast-moving users (dashed lines) in the hierarchical cell structure versus the offered load in each microcell.

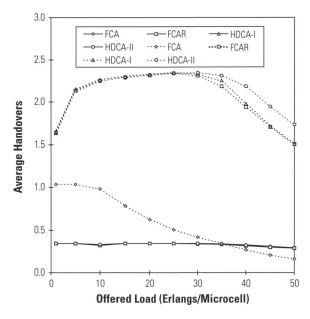

Figure 6.12 Average number of handovers performed by each call in the system versus the offered load in each cell of the first hierarchical level. Bold lines represent slow-moving users and dashed lines fast-moving users.

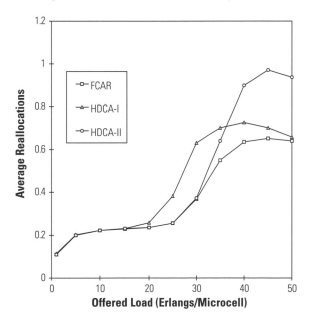

Figure 6.13 Average number of channel reallocations experienced by each call in the system as a function of the offered load in each microcell.

to the new call blocking probability found in the system by fast-moving users (except for FCA), as shown in Figure 6.6. Another factor that contributes to this difference is that the HDP helps to decrease the new call blocking probability for users declared by the system as fast-moving users. Due to the ratio nels for fast-moving users in the macrocell level. In Figure 6.7 the effect of the HDP and overflow are shown clearly. Using the overflow capability in the hierarchical system helps to decrease the new call blocking probability, P_b, for slow-moving users in the system. However, the HDP increases the blocking probability considerably for this type of user at the microcell level (used only by these users) and therefore forces them to overflow into the next hierarchical level. Obviously, FCA is the strategy with the best trade-off for blocking probability in the microlayer and for slow-moving users since it does not use the HDP. Nevertheless, HDCA-II is able to produce the same GOS as FCA for slow-moving users and improves, by far, the behavior of FCA with the other type of user.

HDCA-II is the strategy that produces the lowest overall handover failure probability , P_{hf}, and overall forced termination probability, P_{ft}, (see Figures 6.8 and 6.10). Since in FCA the HDP is not used at all, the macrocell level tends to be saturated by incoming calls from the same level and by calls overflowed by the overlaid level (either handed-up or new calls), increasing the chance of dropping a call in progress of the fast type of user. Meanwhile FCAR, HDCA-I, and HDCA-II present a similar performance for handover failure probability, P_h, for both types of user (Figure 6.9). However, finding a free channel becomes harder after one successful handover, and this is translated into different forced termination probabilities, P_{ft}, for fast- and slow-moving users, as depicted in Figure 6.11.

As expected, the average number of handovers experienced by a slow-moving user is less than that experienced by a fast-moving user. For all the strategies based on HDP, the number of handovers experienced by fast-moving users is almost the same. With the FCA strategy, this number is very small since the number of engaged calls for this type of user is also small. The average number of handovers gives an idea of the signaling load amongst neighboring BSs and the switches involved in the HDP. As can be seen, using HDCA-II, fast-moving users have more handovers on average along the whole evaluation interval due to the larger capacity of HDCA-II translating into a smaller dropping probability. See Figure 6.12.

Figure 6.13 shows another interesting factor of comparison between the strategies considered: the average number of reallocations that a call will suffer in its life. As Figure 6.13 reveals, HDCA-I causes more reallocations only at high teletraffic loads. Under moderate and low loads, it performs as well as FCAR.

6.5.2. Nonuniform Traffic

For this case, the load shown in each cell of Figure 3.4 was varied from 0.2 to 3 times the value shown to take into account nonsymmetric teletraffic loads.

The set of Figures 6.14 to 6.22 compares the strategies studied under nonuniform traffic load. Figure 6.14 shows the overall blocking probability, P_{bt}, versus the load increase. When the offered load is 3 times the load shown in Figure 3.4, the performance of HDCA-II is still better than that experienced with FCA, FCAR, and HDCA-I. HDCA-I performs better than FCAR only for load increases below 2.5 times the load shown in Figure 3.4.

As depicted in Figures 6.15 and 6.16, P_b at both the macrocell and the microcell layers is always smaller when HDCA-II is used. HDCA-I has a good performance also but does not outperform HDCA-II. This shows again that DCA applied at the time slot level improves the system capacity more than when it is applied at the carrier level. As in the case of uniform traffic, the blocking probability, P_b, in the macrocell layer and for fast-moving users is almost the same for the loads considered. Figure 6.16 shows the effect of HDP in the strategies FCAR, HDCA-I, and HDCA-II: different blocking probabilities in the microcell level and for slow-moving users.

As in the case of uniform traffic, the same sorts of comments could be made for the overall probability of handoff failure, P_{ht}, and that experienced by

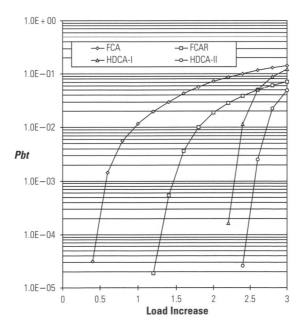

Figure 6.14 Overall blocking probability versus total load variation.

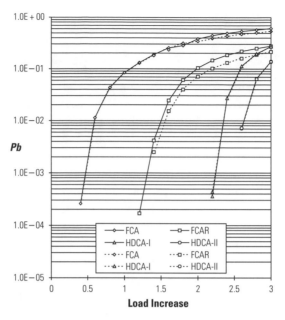

Figure 6.15 Blocking probability in the macrocell layer (bold line) and for fast-moving users (dashed line) in the system as a function of total load variation.

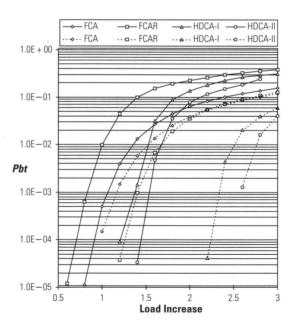

Figure 6.16 Blocking probability in the microcell layer and blocking probability experienced by slow-moving users in the system (bold and dashed lines, respectively) versus total load variation.

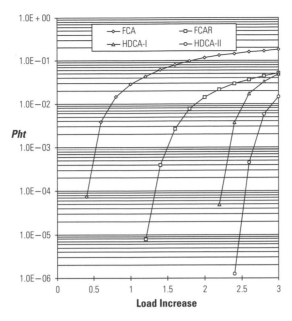

Figure 6.17 Overall probability of handover failure in the system experienced by both types of user as a function of total load variation.

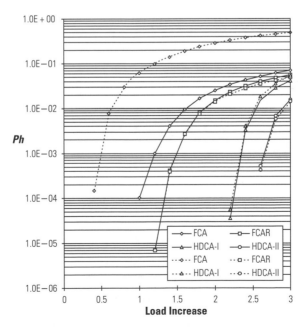

Figure 6.18 Probability of handover failure for slow-moving users and fast-moving users (bold and dashed lines, respectively) versus total load variation.

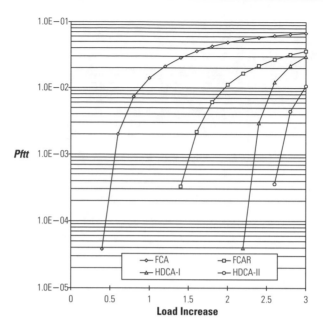

Figure 6.19 Total forced termination probability versus total load variation.

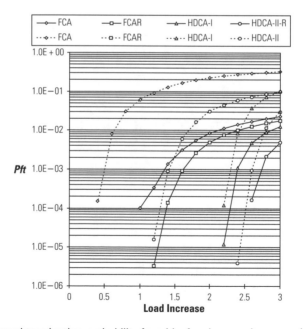

Figure 6.20 Forced termination probability found by for slow-moving users (bold lines) and fast-moving users (dashed lines) in the HCS as a function of total offered load variation.

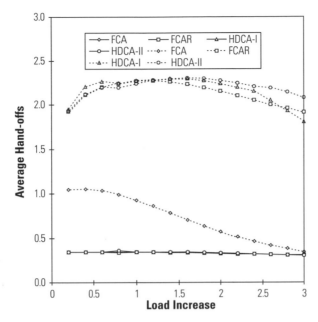

Figure 6.21 Average number of handovers performed by each call in the system as a function of total load variation. Bold lines represent slow-moving users and dashed lines fast-moving users.

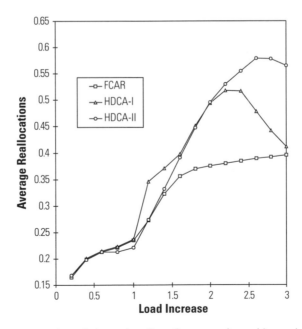

Figure 6.22 Average number of channel reallocations experienced by each call in the system versus the total load variation.

fast- and slow-moving users, P_h, shown in Figures 6.17 and 6.18. HDCA-II considerably decreases the number of calls successfully handed-over even at high teletraffic loads. The other strategies' performances are not even close to that shown by HDCA-II. The preceding comments are also applicable for the total forced termination probability, P_{ft}, shown in Figure 6.19, and the forced termination probability for fast- and slow-moving users, P_{ft}, shown in Figure 6.20. The strategies HDCA-I and HDCA-II keep their excellent balance between P_b, P_{ht}, and P_{ft} even when the load is nonuniform and for high teletraffic loads. Figures 6.14 to 6.18 lead us to affirm that assigning resources at the time slot level is still better than doing it at the carrier level. To take an insight into the average number of handovers done by each type of user in each of the strategies, Figure 6.21 plots this parameter versus the load increase. Due to the reallocations done in HDCA-I, HDCA-II, and FCAR they have similar performances throughout the evaluation range. Obviously, the number of handovers experienced by fast-moving users is greater than that for slow-moving users, but there is no significant difference between HDCA-I, HDCA-II, and FCAR in the number of handovers experienced by a call of any type of user with these strategies. FCA is the strategy where both types of user have the minimum number of handovers due to its higher dropping probability.

It is not surprising than HDCA-II has the largest number of channel reallocations on average in all the evaluation range (Figure 6.22) since the reallocations are done at the time slot level. However, the benefits HDCA-II has shown for P_b, P_h, and P_{ft} and the reduction in signaling load between BSs compensate for the disadvantages that this may imply.

6.6 Notes on the Assumptions Made in This Chapter

Throughout this chapter, several assumptions have been made to simplify the computer simulations and to focus on the performance of the DCA strategies studied. This has made it so that the practical implementation of a HCS has been lost sight of to some extent. Some comments on how to proceed when implementing a HCS in a less homogeneous case are given in this section aswell as the repercussion a different reuse pattern would have on the system.

6.6.1 On Irregular HCS Scenarios

Section 6.4 introduced the evaluation environment used in this chapter. This includes the assumption that each macrocell perfectly covers a homogeneous area of microcells, whereas, in practice, microcells would be placed only in those areas where traffic peaks had been noticed. As a consequence, some macrocells would cover fewer than seven microcells if a reuse pattern of seven is used. The spectrum partitioning would be affected, given the fact that some clusters in the microcell layer would not be complete and regular. Nevertheless, the performance of the DDCA would still be the same and the additional concepts presented here (i.e., hand-downs, overflow, and time slot managing) would still be applicable to a nonregular network.

For this specific case, a more complex approach for spectrum partitioning than that introduced in Chapter 5 would have to be used to take into account that macrocells differ in the number of microcells from which they receive overflow traffic. Obviously, this is not an easy task. A systems operator might find it easy to assign channels to microcells according to demand and based on empirical data or traffic statistics rather than performing a teletraffic analysis.

As for the holes left by absent microcells in irregular scenarios, they would have to be covered by macrocells. All the functions associated with microcells would have to be performed by macrocells. In this sense, the number of hand-ups and hand-downs for slow-moving users performed in the network would increase considerably. The consequence would be an increase of handover signaling between macrocell and microcells involved in hand-downs and hand-ups. The degree to which the number of handovers would increase depends on how irregular the network is.

6.6.2 On the Reuse Pattern

For simplicity the same reuse pattern of seven cells/cluster was used in both layers. This reuse pattern was indeed used in real systems, mainly in first and second generation cellular systems, but its use in UMTS is unlikely because these systems will use a reuse pattern of 1 (or possibly 3 at most). Certainly, in the computer simulations performed in this chapter, different reuse patterns could have been chosen for every layer, giving a different performance for each CAS studied.

The CASs presented here can be implemented in third generation mobile systems as long as the MAS allows it. For example, as they are, HDCA-I and HDCA-II could not be implemented with any MAS proposed for UMTS. However, with some modifications, HDCA-I and HDCA-II could be implemented with the hybrid TDMA/CDMA system proposed for UMTS. HDCA-I would be applied to the time slot level instead of the carrier level and HDCA-II would be applied to the code level rather than to the time slot level.

6.7 Summary

A new DCA strategy, called hierarchical dynamic channel allocation (HDCA) is presented as a solution for increasing demand in third generation systems with HCS. The proposed DCA is evaluated in a HCS using a combination of FDMA/TDMA as the MAS, but the DCA is not restricted to it. This strategy is distributed and the degree of communication amongst BSs is minimum. The solution is easy to implement and reduces the signaling load in comparison with centralized DCA strategies. Also, it is proposed that in a HCS the resource allocator should be in the BS rather than in the central switches, which makes a distributed DCA the best candidate for resource managing. Comparisons with the classic FCA strategy and with an enhanced version of FCA have shown the feasibility of these proposals. HDCA is implemented in two versions: at the carrier level and at the time slot level. Results show that the second option is much better and the additional signaling load cost is small compared with its advantages.

References

[1]   Jolly, W. M., and R. E. Warfield, "Modelling and Analysis of Layered Cellular Mobile Networks," *Teletraffic and Datatraffic in a Period of Change*, Vol. ITC-13, 1991, pp. 161–166.

[2]   Unitel, "Idle Mode Cell Reselection for Microcells, in ETSI GSM2, Ronneby, Sweden, Sept. 1991.

[3]   Yeung, Kwan L., and Sanjiv Nanda, "Channel Management in MIcrocell/Macrocell Cellular Radio Systems," *IEEE Trans. Vehicular Tech.*, Vol. 45, No. 4, Nov. 1996, pp. 601–612.

[4]   Shum, Kenneth W., and Chi Wan Sung, "Fuzzy Layer Selection Method in Hierarchical Cellular Systems," In *Proc. IEEE GLOBECOM'96*, pp. 1049–053, London, UK, Nov. 18–22, 1996.

[5]   Karlsson, Robert, and Jens Zander, "Hierarchical Cell Structures for FRAMES Wideband Wireless Access," In *Proc. ACTS*, pp. 785–791, Spain, 1996.

[6]   Scheibenbogen, Markus, et al., "Dynamical Channel Allocation in Hierarchical Cellular Systems," In *Proc. IEEE Vehicular Technol. Conf. VTC'96*, pp. 721–725, Atlanta, GA, USA, April 28–May 1, 1996.

[7]   Rappaport, S. S., and L. R. Hu, "Microcellular Commun. Systems with Hierarchical Macrocell Overlays: Traffic Performance Models and Analysis," *Proc. IEEE*, Vol. 82, No. 9, Sept. 1994, pp. 1383–1397.

[8]   Ortigoza-Guerrero, Lauro, and A. H. Aghvami, "On the Optimum Spectrum Partitioning in a Microcell/Macrocell Cellular Layout with Overflow," In *Proc. IEEE GLOBECOM 97*, pp. 991–995, Phoenix, AZ, USA, Nov. 1997.

[9]   Jabbari, Bijan, and Woldemar F. Fuhrmann, "Teletraffic Modelling and Analysis of Flexible Hierarchical Cellular Networks with Speed Sensitive Hand-off Strategy," *IEEE Trans. Vehicular Tech.*, Vol. VT-15, No. 8, Oct. 1997, pp. 1539–1548.

[10] Elnoubi, S. M., R. Singh, and S. C. Gupta, "A New Frequency Channel Assignment Algorithm in High Capacity Mobile Communication Systems," *IEEE Trans. Vehicular Tech.*, Vol. VT-31, No. 3, Aug. 1982, pp. 125–131.

[11] West, Kevin A., and Gordon L. Stuber, "An Aggressive Dynamic Channel Assignment Strategy for a Microcellular Environment," *IEEE Trans. Vehicular Tech.*, Vol. 43, No. 4, Nov. 1994, pp. 1027–1038.

[12] Fritz, O., et al., "Comparisons of Different Dynamic Carrier Allocations for Third Generation TDMA Systems," In *Proc. ACTS*, pp. 798–804, Spain, 1996.

[13] Lin, Yi Bing, Anthony R. Noerpel, and Daniel J. Harasty, "The Sub-Rating Channel Assignment Strategy for PCS Hand-Offs," *IEEE Trans. Vehicular Tech.*, Vol. 45, No. 1, Feb. 1996, pp. 122–129.

7

A Distributed Dynamic Resource Allocation Strategy for a Hybrid TD/CDMA System

As mentioned in Chapter 4, TD/CDMA has been selected as one of the MAS to be used for UTRA. In this chapter, an extended DDRA strategy for a HCS based on the strategy HDCA-II presented in the previous chapter is derived and evaluated. The evaluation environment used in this chapter is rather more complex than those used before, and multiple services are considered simultaneously.

In this new version of the DDRA, resources are shared not only between cells of the same hierarchy but between layers of the HCS. In HDCA-II this was not possible. In fact, slow-moving users were only able to use macrocell channels when they were overflowed to the macrocell layer. The DDRA presented here allows one to use macrocell channels in the microcell layer.

This proposed DDRA strategy is evaluated using the hybrid TD/CDMA-FDD proposal made in the Frames Project Mode 1 (FMA1) as a case study. A mixed environment is suggested for the evaluation of the DDRA, which consists of Manhattan-like microcells covered by hexagonal-shaped umbrella cells (macrocells). As in the previous case, users are classified according to their speed as slow-moving users and fast-moving users and are attended by the most suitable layer of the hierarchy according to their speeds. Two types of real time circuit-switched services are considered in the evaluation: speech and data at different rates. The DDRA is compared with the FRA strategy with overflow and with FRA allowing overflow, hand-down, and channel reallocations (FRAHR) (FRA and FRAHR strategies were presented in Chapter 6 and called FCA and FCAR, respectively).

The reason why names have changed from CASs to resource allocation strategies is explained later in this chapter.

7.1 Scope of the Chapter

The main objectives and technical challenges in UMTS are to provide higher spectral efficiency than existing systems and to have resource flexibility to accommodate multiple networks and traffic types within a given frequency band [1]. To reach these goals, WB TD/CDMA has been identified as one of the major MASs for UMTS, other than its main contender, WB-CDMA.

A hybrid TD/CDMA system is based on a TDMA frame structure. The basic way of supporting multiple bit rates is to assign multiple slots to a user. The number of slots inside a frame can be varied dynamically in order to adapt to the transmission of changing service needs. For operation with spreading, multiple codes on a slot can also be assigned. The bit rate for a specific service is fine tuned by selecting an appropriate combination of slot length, burst type, number of slots, and coding rate.

As explained in Chapter 2, FRA is a strategy that does not fully satisfy the requirements of a mobile network, particularly with uneven and time-varying traffic distributions. Therefore, a DRA strategy will certainly be needed to manage all the available resources to increase the capacity of a UMTS network. Some work relevant to DRA strategies can be found in the literature. For instance, in [2] a set of DRA strategies applied to a real GSM network is presented. The main concern of that work is to verify if and to what extent the strategies presented there can offer better performance than FCA in realistic traffic and propagation conditions. In particular, solutions are presented concerning the signaling exchanges entailed by the implementation of each of the DCA strategies considered. In [3], a complete selection of DRA, channel borrowing, and hybrid and reuse partitioning strategies are described. However, their proposals and descriptions are confined to a single layer and no mention is made whatsoever about the implementation of those strategies with HCS or multilayered cellular architectures. Nevertheless, in [4] a DRA for HCS that uses autonomous reuse partitioning is presented. The strategy presented in [4] is called hierarchical autonomous reuse partitioning (HARP). HARP is one of the first dealing with the dynamic allocation of channels in a HCS. It takes into account the specific situation of a HCS and offers the possibility of controlling the capacity between layers. Due to its nature, this algorithm is completely decentralized but is able to estimate the current load and the free capacity of the different layers. In HARP channel ordering is used. The first channels of the ordered list are used by the microcells, whereas the last channels in the list are used by the macrocells. In this way, only the channels required by each layer are used and the rest are left for the other layer. When there are no resources available, new calls are blocked from the system. This allows a flexible separation of channels between layers that can adapt according to their needs. The HARP algorithm is compared with two other algorithms. The first of them does not differentiate the layers and therefore no channel ordering is used. Micro- and macrocells can use any channels at any time as long as they are available. In the second algorithm there is a channel separation between layers, but this spectrum partitioning is fixed, similar to that studied in Chapter 5 and applied to the network proposed in Chapter 6.

In [5], a CAS that uses hand-down and hand-up is presented. It is called combined channel assignment (CCA) and is, so far, the most complex strategy for HCS described in the literature; but there is no dynamic assignment of resources either within the layers or between layers. The CCA strategy combines overflow, underflow, and reversible techniques. When new or handover calls have no available channel to use in the overlaid microcell they can overflow to use free channels in the overlaying macrocell. Handover calls can underflow to use a free channel in the microcell, and handover attempts from overlaying macrocell to overlaid microcell are reversed to use free channels in the microcell if there are idle channels in the microcell.

This chapter describes a DDRA strategy for a HCS that can be applied to a hybrid TD/CDMA system, to a CDMA system (WB CDMA with TDD) or to a pure TDMA system. The content of this chapter is the extension and evaluation by computer simulation of a DDRA strategy suitable for UMTS. The dynamic nature of this strategy permits its adaptation to uneven and changing traffic, while the distribution of the decision-making process between the cells reduces the required computation and communication between BSs. This strategy is evaluated using FM1 [1, 6] in a practical HCS that could appear in future UMTS. Two types of circuit-switched service are considered in the evaluation: voice at 8 Kbps and data transmissions at 32 Kbps, 64 Kbps, 144 Kbps, and 384 Kbps. Also, an analytical method is presented to assess the blocking probability in a HCS using FRA with different types of user and service.

7.2 System Description

We start with the definition of the basic concepts that are used throughout the chapter and the operations involved in the allocation strategies.

7.2.1 Resources

In a hybrid WB TD/CDMA system users are separated orthogonally into time slots, and within each time slot an additional separation by spreading codes is used. The method to be used depends on the service and radio conditions handled by the link adaptation. The unit of transmission is one slot. This unit is then divided into smaller units, either subslots or by spreading codes if the spreading feature is being used. In this way, a smaller granularity is achieved. Given that a user might require several resources, a DRA scheme could assign several spreading codes to the same time slot (multiple code option), several channels in different time slots (multiple slot option), or a combination thereof. In this chapter, it is proposed, moreover, that several time slots/codes from different carriers may be combined to deal with a given type of service. The resource allocation strategies presented will be applied to a portion of a time slot or to a code associated with a time slot, and these will hereafter be called simply a resource. The goal of the allocation strategy is that all available time slots are used with approximately the same number of active spreading codes [6].

7.2.2 Description of the HCS Environment

Figure 7.1 depicts a large geographical area covered by a set of contiguous Manhattan-like microcells. Every W microcells are overlaid by one large hexagonal-shaped macrocell only, which fulfills the role of umbrella cell. The microcells constitute the lower layer, and the macrocells form the upper layer of a two-layer HCS. It is assumed that there is a large number of users randomly traversing the area covered, each of them with a different speed. Users are classified as fast- and slow-moving users, and they do not change their speed class during a service. They are tracked only in the microcellular layer since this can give all the information required such as when to initiate a handover and to which neighboring cell it is moving. Each layer in the network is given a subset of physical channels by the spectrum partitioning strategy [7] derived from the total available spectrum. Then, every cell in the HCS is assigned a set of Y resources. All physical channels are shared among new calls and handover calls; that is, no prioritization of handover calls by means of reserved physical channels is used. No resource reservation applies for any of the different types of service available in the system; hence, all of them can use any of the resources as long as they are available.

The HCS proposed is a regular grid of microcells and macrocells that may not exist in real systems. A uniform structure is proposed because this will simplify the teletraffic analysis presented here. Systems operators may want to

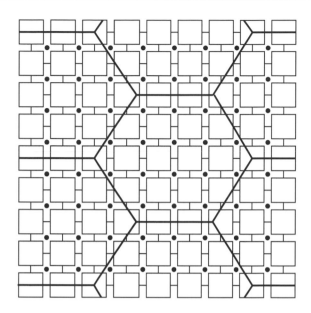

Figure 7.1 Evaluation environment. ©1999 IEEE. Reprinted, with permission, from [8].

know how an irregular or a more disjointed system would behave in reality. The explanation given in the last section of Chapter 6 applies to this case.

7.2.3 Frames Proposal Mode 1

Within the framework of the advanced communication technologies and services (ACTS) program, the project FRAMES is in charge of defining a MAS for third generation UMTS, based on adaptive radio interface concepts. The evaluation has led to the FMA concept, which has been presented to ETSI for consideration in the UMTS standardization process. Two multiple access techniques have been identified: WB-TD/CDMA-TDD and WB-CDMA, which can both meet the UMTS requirements. They are based on the FRAMES project mode 1 (FM1) and mode 2 (FM2), respectively [1, 6, 9]. These modes are harmonized with each other and also support compatibility with GSM. Recently (January 1999), WB-CDMA-FDD and WB-TD/CDMA-TDD were chosen for operation in paired and unpaired spectrum, respectively. FM1, with spreading, facilitates the application of a variety of DCA strategies, where the allocation of channels depends on the current traffic load and/or the current interference conditions [10]. The evaluation of the proposed resource allocation strategies presented here is done in a system using the FM1 proposal.

7.3   Description of Strategies Considered

Three different strategies are considered: classic FRA with overflow procedure, FRA with resource reassignments and hand-down procedure (FRAHR), and DDRA. In the following, we describe the operation of each of them.

7.3.1   FRA

The FRA strategy involves the following system operations:

- New requests for service and new handover requests (voice or data transmission) originated by a slow-moving user will be directed toward the preferred microcell (a preferred cell in each layer is that cell that receives the strongest signal from the mobile in the particular HCS level). They will be attended if there are enough idle resources; otherwise they will be overflowed to the preferred macrocell. The overflowed requests will be accepted by the macrocell if the number of idle resources is equal to or larger than the required number; otherwise the call will be either blocked or dropped (forced to terminate), whichever is applicable.

- A new request for service originated by a fast-moving user is directed to its preferred macrocell and will be accepted if there are enough idle resources to serve it; otherwise, the call will be blocked. When handover is performed either by a fast-moving user or by a slow-moving user in the macrocell layer, the target macrocell will try to accommodate the handover request assigning enough resources. If the number of available resources is insufficient, the call will be dropped.

7.3.2   FRA with Hand-Down Procedure and Resource Reassignments

Since the resources in the macrocell layer are scarce, the HDP described in the previous chapter has been added to the FRA strategy to increase the possibility of successful resource acquisition by fast-moving users in the macrocell layer. The way HDP is applied to FRA to create FRA-HR is described as follows.

- *Resource Acquisition*: New requests for service and hand-off requests originated by a fast-moving user in a macrocell will be attended as long as there are enough idle resources. When this is not the case, HDP takes place (provided that there is at least one slow-moving user being attended in the macrocell). When a single execution of the HDP does not release the required number of resources, it may be repeated more than once to get the desired resources. If there are several slow-moving users being attended by the macrocell, the HDP will select that slow-moving user that would put into the idle state a number of resources equal to or closest to the desired number. Note, the more microcells that are overlaid by the macrocell, the more the chances of the HDP being successful. This same process is applied when requests originated by slow-moving users overflow from the microcell layer (either new requests or handover requests) to the macrocell layer.

- *Release of Resources*: When a slow-moving user finishes its connection to the network in a microcell, all the resources it was occupying are released. A set of candidate slow-moving users being attended to by the overlaying macrocell is then found. If the number of resources just released are enough to take back one of these slow-moving users from the macrocell layer, HDP is performed.

7.3.3 DDRA

The DDRA strategy presented here and applied to a HCS is a generalization of the HDCA-II presented in Chapter 6, modified, and complemented to account for HCS scenarios, the different types of service offered, and users with different speeds. As in the case of HDCA-I and HDCA-II, what makes the DDRA scheme suitable for use in a HCS is the fact that the resource allocation and the release policy for both FRA and FRA-HR are used in this strategy. Important features added to the DDRA are the capabilities of sharing resources between layers and of actually changing the spectrum partitioning according to different or time-varying conditions of traffic (e.g., propagation and traffic jams).

When a request for service arrives at a cell in a particular layer of the HCS, the DDRA gets the (or part of the) required resources from that cell. If this procedure does not satisfy the users requirements, then the DDRA finds the (part of the, or the rest of the) required resources by:

- Borrowing a channel (according to SDCA) in the same hierarchical level;
- Applying the HDP;
- Borrowing resources from the upper hierarchical layer.

The third option deals with spectrum sharing between layers in the HCS and can be seen as the ultimate step in the allocation of resources by a call and is avoided whenever possible to minimize blocked calls in the upper layers. In a two-layered-HCS, the use of borrowed resources from a lending macrocell will be forbidden in the macrocell layer while they are being used in the microcell layer; as soon as they are released, they will be returned to the original macrocell. Note that the use of resources borrowed from a macrocell in a microcell does not imply restricting their use in cochannel macrocells because the reuse constraints will still be fulfilled.

7.4 Evaluation Environment

The exact conditions used in our computer simulations are described here. A two-layered environment consisting of microcell and macrocell layers is assumed. The microcellular environment is formed of a (regular) rectangular grid of intersecting streets (referred to as Manhattan-like microcells), as shown in Figure 7.1. The BSs are at street lamp height and are placed at the center of each crossroad (clover leaf cells). In a real system, assuming that the proper antenna arrays are used to ensure continuity of RF signals round the corners, this would ease the problems related to handover, since users do not experience a sudden drop of signal when turning at a street corner [11]. BSs are separated by 200m. Users are only tracked at the microcell level and a wraparound topology is used so that the boundary effects in the simulations are eliminated, as depicted in Figure 7.2. The interference neighborhoods of each cell also wrap around. The macrocell layer is formed of hexagonal cells overlaying the microcells, and their BS antenna height is above the average roof top height. According to [1], a cluster size in the range of 3 to 9 has to be supported for FRAMES1. In our simulations, a reuse pattern of four was used for both the microcell and the macrocell layers, as depicted in Figure 7.3(a).

Note that in microcellular structures with Manhattan-like cells reuse patterns may not be as symmetrical as in hexagonal-shaped cells scenarios and practical implementations may differ from theoretical approaches (e.g., theoretically in Manhattan-like structures the reuse pattern of three does not exist but it can be implemented practically. In fact, it is being proposed for UMTS) and they also depend on the separation and location of BSs. For Manhattan-like cells, theoretically only the patterns given by $N = i^*i + j^*j$ can be deployed (i.e., $N = 1, 2, 4, 5, 8, 9, 10, 13, \ldots$). Figure 7.3(a) shows a reuse pattern in which every different set of channels is reused both horizontally and vertically every four cells. This reuse pattern has previously been proposed in the literature and used to simulate a microcellular network with Manhattan-like cells. Diagonally the set of channels is reused immediately, but if we consider that there are no reuse constraints on NLOS cochannel cells because of the large signal attenuation around the corner, then we can consider this reuse pattern as

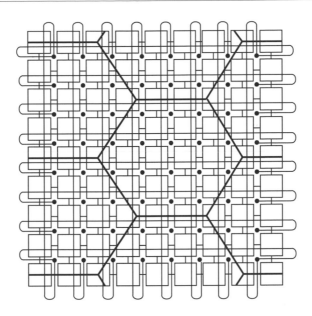

Figure 7.2   Wraparound topology. ©1999 IEEE. Reprinted, with permission, from [8].

a four reuse pattern, although, strictly speaking, it is not the most correct way of doing it because it does not have the properties of symmetry and compactness. Figure 3(b) shows a symmetric and compact reuse pattern of four.

1	2	3	4	1	2	3	4
4	1	2	3	4	1	2	3
3	4	1	2	3	4	1	2
2	3	4	1	2	3	4	1
1	2	3	4	1	2	3	4
4	1	2	3	4	1	2	3
3	4	1	2	3	4	1	2
2	3	4	1	2	3	4	1

(a)

1	2	1	2	1	2	1	2
3	4	3	4	3	4	3	4
1	2	1	2	1	2	1	2
3	4	3	4	3	4	3	4
1	2	1	2	1	2	1	2
3	4	3	4	3	4	3	4
1	2	1	2	1	2	1	2
3	4	3	4	3	4	3	4

(b)

Figure 7.3   Cell reuse pattern for the microcellular layer: (a) practical approach, (b) theoretical approach.

Six carriers of 1.6 MHz were assumed to be available in a FDD approach: three carriers for the uplink and three more for the downlink. Hence, the total number of 1/64 time slots or codes available is 192 (we assume that up to eight bursts can be transmitted within one time slot, even if the bursts are allocated to a single user). An approximate partitioning of 60% of resources for the microcell layer and 40% resources for the macrocell layer is used [7]. As a result, a total of 28 1/64 time slots or codes are given to each microcell and a total of 20 to each macrocell in the system. Five types of real time service are considered. Speech service is available at 8 Kbps as well as data transmissions at 32 Kbps, 64 Kbps, 144 Kbps, and 384 Kbps. All the services are circuit switched with 100% activity. Users are classified as fast- and slow-moving users, as described in previous sections, with a probability of a user being of the slow type equal to 0.9 and therefore, with probability 0.1 of being a fast-moving user. This is an arbitrary ratio that may seem fairly high considering that to have such amount of microcells covered by each macrocell implies a high dense city center-type environment where even the cars move slowly.

Assuming that the required number of physical channels (codes/time slots) per frame is 0.5, 2, 4, 9, and 24 for the type of service 1, 2, 3, 4, and 5, respectively, and considering that there are 64 physical channels per frame, Table 7.1 shows the types of service considered along with all their associated requirements.

In general, the models used with this evaluation environment are different from those used within the ETSI SMG 2 evaluation process [12]. Hence, the results presented here may not be the same as those obtained according to [12].

The mobility model adopted in this chapter is exactly the same as that described in Section 4.3.2.

Table 7.1
Service Classes and Examples for Mixed Services for FMA1 with Spreading.
©1999 IEEE. Reprinted, with permission, from [8].

Parameter	Type of service				
	1	2	3	4	5
Service (Kbps)	8	32	64	144	384
Maximum traffic per carrier (Erlangs)	128	32	16	7	2
Code rate	0.55	0.55	0.55	0.55	0.55
Modulation	QPSK	QPSK	QPSK	QPSK	QPSK

7.4.2 Propagation Model for the Microcellular Layer

The propagation model used for the microcellular layer was originally presented in [13] and is fully described in Section 4.3.3.

7.4.3 Propagation Model for the Macrocellular Layer

For the macrocell layer, the path losses are calculated according to the formula [12]

$$L = -10 \log_{10} \left[\left(\frac{\lambda}{4\pi R} \right)^2 \right] - 10 \log_{10} \left[\frac{\lambda}{2\pi^2 r} \left(\frac{1}{\theta} - \frac{1}{2\pi + \theta} \right)^2 \right] \quad (7.1)$$

$$- 10 \log_{10} \left[(2.35)^2 \left(\Delta h_b \sqrt{\frac{d}{\lambda}} \right)^{1.8} / R^{2(1-0.004\,\Delta h_b)} \right]$$

where

$$\theta = \tan^{-1} \left(|\Delta h_m| / x \right) \quad (7.2)$$

$$r = \sqrt{(\Delta h_m)^2 + x^2}$$

Δh_m is the difference between the mean building height and the mobile antenna height and x is the horizontal distance between the mobile and the diffracting edges. In the practical case when Δh_m = 10.5m and x = 15m, which is typical in an urban and suburban environment with an average building height of four stories, the path loss expression reduces to [12]

$$L = \left[40\left(1 - 0.004\,\Delta h_b \right) \right] \log_{10}(R) - 18 \log_{10}(\Delta h_b) + 21 \log_{10}(f) + 80 dB \quad (7.3)$$

where R is the distance from the transmitter to the receiver in kilometers, Δh_m is the BS antenna height measured from the rooftop, and f is the frequency in megahertz. The shadowing effect is modeled as $10^{\xi/10}$, where ξ is a Gaussian variable with zero mean and standard deviation of 6 dB.

7.4.4 CIR Threshold

As explained in Chapter 4, only the DL CIR is calculated since it is assumed that this will be worse than the UL and thus provide the limit to system performance. The way the DL CIR is calculated and mapped to the Eb/N_0 is described in Section 4.3.7. Once the CIR is calculated accordingly, this ratio is matched to the appropriated Eb/N_0, as specified in Table 7.2 (a more sophisticated simulation should take into account different thresholds for each different data transmission).

7.4.5 Teletraffic Model

As already stated, users are only tracked in the microcell layer. This allows us to once again use the active-dormant Markov model, introduced in Section 4.3.4, in our simulations to account for even and uneven teletraffic distributions in the proposed scenario. Some of the parameters used in the simulations are shown in the Table 7.3 [14]. Most of them are related to the active-dormant Markov model.

The total offered load, $a_{i,T}$, to each microcell i is given by

$$a_{i,T} = a_{i,1} + a_{i,2} + \cdots + a_{i,5} = \sum_{j=1}^{5} a_{i,j} \qquad (7.4)$$

where $a_{i,1}$ to $a_{i,5}$ are the offered loads produced by each of the five types of service available and are given by

$$a_{i,j} = a_{i,T} f_j \qquad (7.5)$$

where f_j is the proportion of the total offered load submitted to microcell i by type j users (see second row of Table 7.3).

Table 7.2
Eb/N$_0$ Thresholds for the Multiple Services in the CDMA/TDMA System.
©1999 IEEE. Reprinted, with permission, from [8].

Type of Service	Average Eb/N$_0$ (dB)	
	Macrocell	Microcell
Spread speech @BER=10^{-3}	10	12
Spread data @ BER=10^{-5}	15	17
Non-spread speech @ BER=10^{-3}	11.5	13.5
Non-spread data @ BER=10^{-5}	8.5	11

Table 7.3
Traffic Parameters Used in the Simulation
©1999 IEEE. Reprinted, with permission, from [8].

Type of service	1	2	3	4	5
Average service duration (min)	1.66	0.035	0.017	0.0592	0.022
Proportion of the total offered load	0.5	0.249	0.2	0.05	0.000025
Offered traffic load per microcell (Erlangs)			0–15		
Active to dormant traffic ratio			5		
Average duration of DORMANT mode (min)			10		
Average duration of ACTIVE mode (min)			1		
Average No. of handovers per call for *fast moving users*			2.0		
Average No. of handovers per call for *slow moving users*			0.28		
No. of codes (1/64Time Slots) per microcell/macrocell (FRA)			28/20		
No. of codes/time slots per 1/8 Slot			8		
Division multiplex technique			FDD		
No. of 1.6 MHz carriers			6		
Average No. of microcells overlaid per macrocell			16		
Total No. of macrocells			5		
Total No. of microcells			64		

7.5 Mathematical Approach to Analyze the HCS

In this section, an analytical method for assessing the blocking probability is presented for each type of user in the hierarchical network using pure FRA. Neither hand-up nor hand-down is considered. Expressions for the handover failure probability are found as well. The system is assumed to be homogeneous, so all cells in the same hierarchical level are statistically identical. In the equilibrium state, the overall system can be analyzed by focusing on only one cell in each level [14]. Let us define an area Ω consisting of one macrocell overlaying W (i = 1, 2, … , W) Manhattan-like microcells. The total offered load to the area is given by

$$A_\Omega = Wa_\mu + a_M \tag{7.6}$$

where a_μ is the offered load submitted to a single microcell and a_M is the offered load submitted to the overlaying macrocell. From the total offered load to an area, a fraction G_μ is sent to the microcell layer and a fraction G_M is sent to the macrocell. So we can write the offered load to a single microcell a_μ and to the macrocell a_M as

$$a_\mu = G_\mu A_\Omega / W \tag{7.7}$$
$$a_M = G_M A_\Omega$$

where G_μ = 0.9 and G_M = 0.1. G_μ and G_M could be understood as the probability that the speed of users is, respectively, smaller or larger than a predefined speed threshold to classify users as slow- or fast-moving users (note also that $a_\mu = G_\mu a_{i,T}$, with $a_{i,T}$ as defined in the previous section if uniform traffic is considered). The following memoryless assumptions allow the problem to be cast in the framework of a multidimensional birth and death process [15]. It should always be borne in mind that there are five types of service (one voice and four data) in the system under consideration.

1. New call arrival processes offered to a given cell are Poisson processes for all five types of service. The mean new call arrival rates from slow-moving users to each microcell for each type of service are λ_{S1}, λ_{S2}, λ_{S3}, λ_{S4}, and λ_{S5}; and the mean new call arrival rates from fast-moving users to each macrocell for each type of service are given by λ_{F1}, λ_{F2}, λ_{F3}, λ_{F4}, and λ_{F5}.

2. The handover call arrival processes due to the motion of slow-moving users in microcells are also considered to be Poisson processes with mean handover arrival rates in the next microcell λ_{Sh1}, λ_{Sh2}, λ_{Sh3}, λ_{Sh4}, and λ_{Sh5}, respectively, for each type of service and in the macrocell layer with mean handover arrival rates λ_{SH1}, λ_{SH2}, λ_{SH3}, λ_{SH4}, and λ_{SH5}, respectively. The handover arrival processes due to the motion of fast-moving users in the macrocell layer are also considered to be Poisson processes with mean handover arrival rates λ_{FH1}, λ_{FH2}, λ_{FH3}, λ_{FH4}, and λ_{FH5}, respectively, for each type of service.

3. The calls from either slow- or fast-moving users have an unencumbered call duration according to a negative exponential distribution with parameters $\mu_1, \mu_2, \mu_3, \mu_4$, and μ_5.

4. The cell dwell time, the time spent by a MS in a cell independent of being engaged in a call, is a random variable approximated by a negative exponential pdf (clearly, this assumption would not be valid for fast-moving users in microcells but is valid for slow-moving users). For slow-moving users traveling across a microcell, the cell dwell time has mean equal to η_{S1} and, while traveling in a macrocell, a mean equal to η_{S2}. On the other hand, fast-moving users traveling in a macrocell have a cell dwell time with a mean equal to η_{F2}.

7.5.1 Microcell Layer

In a birth and death process the probability of a unique transition in a very small interval of time is directly related to the interval duration, while the probability of a transition involving more than one state in a very small interval of time is equal to zero. This means that, in a given time, there can only be a birth, a death, or the possibility of remaining in the same state. Hence, there can only be transitions to neighboring states. The case where a user needs more than one channel to establish a connection cannot be solved directly by matching a birth and a death with the occupation or release of a single channel in the system but with a transition to the state produced by the arrival or departure of a certain type of user.

The transitions produced by a birth or a death will be made to a neighboring state when the user requires only one channel, but nonneighboring states will be reached when more than one channel is required by the user. However, if we represent the number of users in the transition diagram instead of the number of busy channels in the system, an exact solution could be found as long as the number of users occupy less than or equal to the number of available channels. An example of a state transition diagram is shown in Figure 7.4 for the case of two types of user, with one and two channels required per connection respectively.

Since there are five types of service available in the system, a_μ is formed of five different streams of traffic produced by slow-moving users and, therefore, can be expressed as

$$a_\mu = a_{\mu 1} + a_{\mu 2} + \cdots + a_{\mu 5} = \sum_{i=1}^{5} a_{\mu i} \qquad (7.8)$$

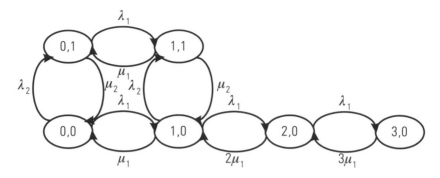

Figure 7.4 Possible state transitions diagram for the case where there are three resources available for two different types of user. One of them requires one resource for the communication and the other requires two (represented by the first and the second of the pair of numbers in each state, respectively). ©1999 IEEE. Reprinted, with permission, from [8].

with

$$a_{\mu i} = \left(\lambda_{Si} + \lambda_{Shi} \right) / \left(\mu_i + \eta_{S1} \right) \tag{7.9}$$

Every microcell has S channels with no guarded channels either for handover calls or for new calls of any type of service. Clearly, this system should be solved by an X-dimensional birth and death process where X is equal to the number of services available, in this case 5. The process for drawing the equilibrium state equations is obviously a tedious task, even when each type of user makes use of only one channel per connection. However, since each type of user is independent, and since each type of service is requested independently, the joint probability that there is a combination of S or fewer users of any type in the system can be found as the multiplication of each of the marginal distributions. That is, each marginal distribution is given by

$$P\left(j_i \right) = \left[\left(\lambda_{Si} + \lambda_{Shi} \right) / \left(\mu + \eta_{S1} \right) \right]^{j_i} c_i \, / \, j_i! = a_{\mu i}^{j_i} c_i \, / \, j_i! \tag{7.10}$$

with $i = 1, 2, \ldots, X$, where X is the number of services available and c_i is a constant. Then, the probability that the system is in the state $P(j_1, j_2, \ldots, j_X)$ is given by

$$P\left(j_1, j_2, \ldots, j_X \right) = \prod_{i=1}^{X} \frac{a_{\mu i}^{j_i}}{j_i!} c_i = \prod_{i=1}^{X} \frac{a_{\mu i}^{j_i}}{j_i!} P_0 = \frac{a_{\mu 1}^{j_1} \, a_{\mu 2}^{j_2} \cdots a_{\mu X}^{j_X}}{j_1! \, j_2! \cdots j_X!} P_0 \tag{7.11}$$

remembering that

$$\sum_{j_1=0}^{S/m_1}\sum_{j_2=0}^{S/m_2}\cdots\sum_{j_X=0}^{S/m_X}P\big(j_1,j_2,\ldots,j_X\big)=1 \tag{7.12}$$

where m_1 to m_X represent the number of channels required by each type of service. Then

$$P_0=P(0,0,\ldots,0)=\left[\sum_{j_1=0}^{S/m_1}\sum_{j_2=0}^{S/m_2}\cdots\sum_{j_X=0}^{S/m_X}P\big(j_1,j_2,\ldots,j_X\big)\right]^{-1} \tag{7.13}$$

where S/m_1 to S/m_X should be understood as the integer part of S/m. Finally,

$$P\big(j_1,j_2,\ldots,j_X\big)=\frac{\displaystyle\prod_{i=1}^{X}\frac{a_{\mu i}^{j_i}}{j_i!}}{\left[\displaystyle\sum_{g_1=0}^{S/m_1}\sum_{g_2=0}^{S/m_2}\cdots\sum_{g_X=0}^{S/m_X}\left(\prod_{i=1}^{X}\frac{a_{\mu i}^{j_i}}{g_i!}\right)\right]} \tag{7.14}$$

subject to the very important condition $m_1 g_1 + m_2 g_2 + \cdots + m_X g_X \le S$.

The blocking probability for each type of user is given by the summation of all the valid states for which the condition $S - k \le m_i$ is fulfilled. This is

$$pb_i=\sum_{S-k<m_i}P\big(j_1,j_2,\ldots,j_X\big) \tag{7.15}$$

with $k = m_1 g_1 + m_2 g_2 + \cdots + m_X g_X$. The overall blocking probability in the microcell layer is then given by

$$pb_T=\sum_{i=1}^{X}pb_i\,\frac{a_{\mu i}}{a_\mu} \tag{7.16}$$

Since there is no channel reservation, the handover failure probability for each type of user, ph_i, is equal to the blocking probability of each type of user, so $ph_i = pb_i$. The same can be said about the overall handover failure probability; hence, $ph_T = pb_T$. Beginning with an initial estimate of handover arrival rates for each of the services available, the equations are solved for the state probabilities, which are then used to determine the average handover departure rates for each type of service in much the same way as described in [16]. A big difference here is that five types of handover arrival rates are being considered in the calculation.

7.5.2 Overflow Traffic and Macrocell Layer

The composite overflow traffic is modeled by an IPP. The way to determine the IPP parameters $(\lambda, \gamma, \omega)$ can be found in [15]. λ represents the intensity of the modulated Poisson process, $1/\gamma$ is the mean on-time, and $1/\omega$ is the mean off-time of the random switch that modulates the IPP. There are 10 different streams overflowing from the microcell layer to the macrocell layer: they are new and handover calls from five type of services produced by slow-moving users only. The on-time and off-time random variables follow a negative exponential probability density distribution. Once the parameters are determined, the states equations can be easily formulated in the usual way [15]. This process is rather complicated because there are also fast-moving users in the macrocell layer. Appendix A shows the derivation of the state equations to calculate the blocking probability in the system for every type of service. The interested reader is also referred to [15].

A numerical example of the application of this analytical method under uniform traffic is given in Figure 7.5 when applied to a system like that described in previous sections. η_{S1}, η_{S2}, and η_{F2} are set to 5.83 min, 15.43 min, and 2.2 min, respectively. The blocking probability for each type of user is plotted against the total offered load per microcell. As can be seen, there is close agreement between the numerical and the simulation results (dashed and solid lines, respectively).

7.6 Examples And Discussion

This section presents simulation results generated for the FRA, FRAHR, and the DDRA strategies. The purpose of these simulations is to compare the performances of FRA, FRAHR, and DDRA for different types of service in a HCS. All the simulation results are applicable to the FM1 proposal for UTRA (see Section 4.1).

Figures 7.6 to 7.11 compare the probability of new call (or data transmission) blocking for the three strategies when plotted versus the offered load per microcell. Figure 7.6 shows a substantial reduction in the overall new call blocking probability using the DDRA strategy in comparison with the two

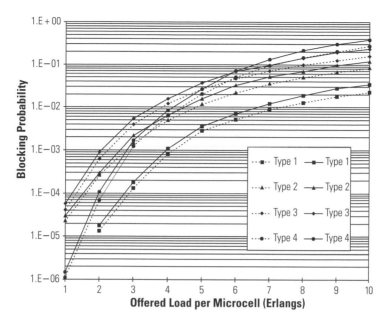

Figure 7.5 Blocking probability for each type of service in a HCS with different types of user and different types of service using the method described in Section 7.5 (solid lines denote simulation and dashed lines denote numerical results).©1999 IEEE. Reprinted, with permission, from [8].

other strategies. DDRA clearly outperforms FRA and FRAHR in all the evaluation range in terms of the overall blocking probability for each type of service, as shown in the set of Figures 7.7 to 7.11. For all the cases shown, with a fixed blocking probability of 1%, the DDRA strategy always presents an increase in capacity of at least 30% in comparison with the FRAHR strategy. Using a DRA rather than a FRA strategy a smaller blocking probability is experienced in the HCS for new calls from all types of service, both for slow- and for fast-moving users. It can also be seen that, even when the offered load increases, the advantages that the FRAHR strategy has in comparison with the FRA strategy are almost constant. This means that using the HDP always gives the FRA strategy capacity gain for all kinds of traffic loads (light and heavy). As Figure 7.10 shows, the service at 144 Kbps has good performance and the system can cope with it without major problems. However, the data service at 384 Kbps experiences a very high blocking probability (see Figure 7.11) due to the limited number of resources in the microcell layer, and in the system in general. The system requires the availability of more resources to improve this service's

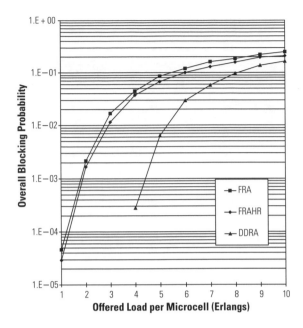

Figure 7.6 Overall blocking probability in the HCS. ©1999 IEEE. Reprinted, with permission, from [8].

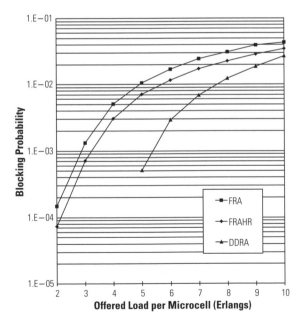

Figure 7.7 Blocking probability in the HCS for a user with a service of type 1 (speech). ©1999 IEEE. Reprinted, with permission, from [8].

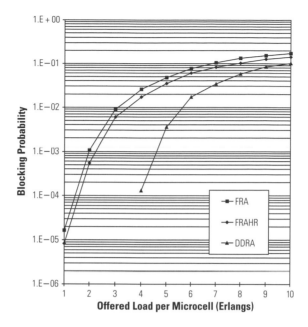

Figure 7.8 Blocking probability in the HCS for a user with a service of type 2 (data at 32 Kbps). ©1999 IEEE. Reprinted, with permission, from [8].

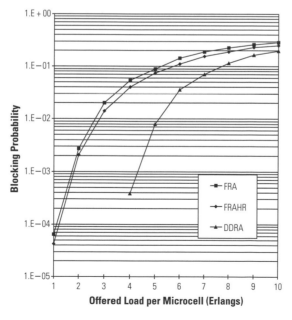

Figure 7.9 Blocking probability in the HCS for a user with service of type 3 (data at 64 Kbps). ©1999 IEEE. Reprinted, with permission, from [8].

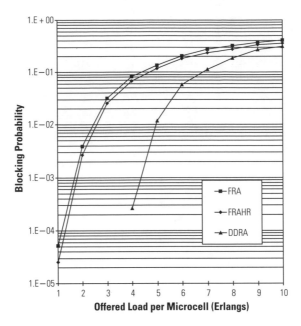

Figure 7.10 Blocking probability in the HCS for a user with a service of type 4 (data at 144 Kbps). ©1999 IEEE. Reprinted, with permission, from [8].

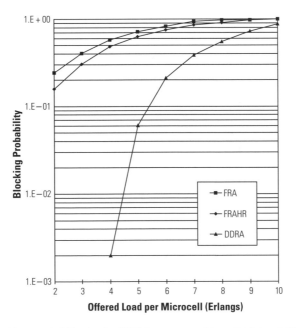

Figure 7.11 Blocking probability in the HCS for a user of type 5 (data at 384 Kbps). ©1999 IEEE. Reprinted, with permission, from [8].

performance. From this part of the evaluation it is observed that DDRA is the strategy that makes the best balance between channel usage in the macrocells and their role as overflow channels. Sharing channels not only between cells of the same hierarchy but between cells of different layers, makes the DDRA strategy a powerful scheme to increase capacity in the HCS.

On the other hand, Figures 7.12 to 7.16 show and compare the forced termination probability for the FRA, the FRAHR, and the DDRA strategies. Observe from Figure 7.12 that a considerable reduction in the overall forced termination probability of the system is achieved when DDRA is used rather than a FRA-based strategy. Very small forced termination probabilities can be expected for all types of service with a DRA, as shown in the set of Figures 7.12 to 7.16, where the total forced termination probability for each type of service (except type 5) is plotted against the offered load in each microcell. As seen in this set of graphs, the advantages of using the DDRA strategy are kept throughout the evaluation range and the DDRA strategy performs better than the FRA strategies. The HDP always helps the FRAHR strategy to decrease the number of dropped calls throughout the evaluation range when compared to the FRA strategy. The forced termination probability experienced by fast-moving users with types of service 1 to 4, when the DDRA is being used in the system is nil.

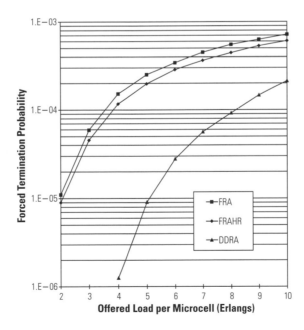

Figure 7.12 Overall forced termination probability. ©1999 IEEE. Reprinted, with permission, from [8].

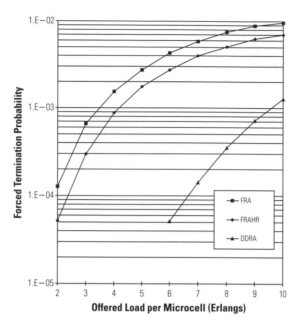

Figure 7.13 Forced termination probability for users with a service of type 1 in the HCS. ©1999 IEEE. Reprinted, with permission, from [8].

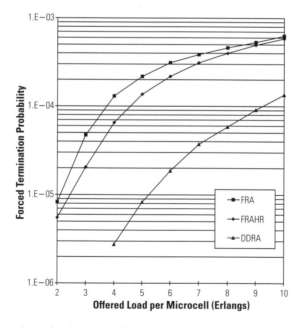

Figure 7.14 Forced termination probability for users of type 2 in the HCS. ©1999 IEEE. Reprinted, with permission, from [8].

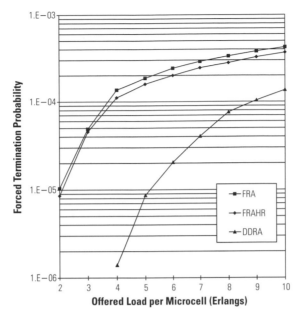

Figure 7.15 Forced termination probability for users with service of type 3 in the HCS.
©1999 IEEE. Reprinted, with permission, from [8].

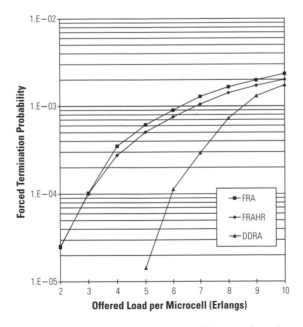

Figure 7.16 Forced termination probability for users with a service of type 4 in the HCS.
©1999 IEEE. Reprinted, with permission, from [8].

Therefore, the forced termination probability shown in Figures 7.13 to 7.16 for the DDRA strategy is that experienced by slow-moving users only. Due to a high blocking probability for the type of service of data transmission at 384 Kbps, the number of users engaged in a conversation in the HCS is very small; and the forced termination probability for this type of service is also nil, and so no graph is given.

From the system operator point of view, it is important to be given some figures that could affect the signaling load of the system. The number of handovers in the lifetime of a conversation is an important parameter to consider. Figure 7.17 shows the average number of handovers necessary in the lifetime of a conversation for slow- and fast-moving users (for any type of service). Clearly, under the DDRA strategy, this number is large compared to the FRA and FRAHR strategies, but there is an obvious trade-off with the number of blocked calls, particularly in the case of fast-moving users. Finally, in Figure 7.18 the average number of reallocations that a call suffers is presented. Obviously, FRAHR has the smallest number of reallocations, as compared to DDRA; nevertheless, it is worth having this increase in the system signaling load because of the advantages already explained.

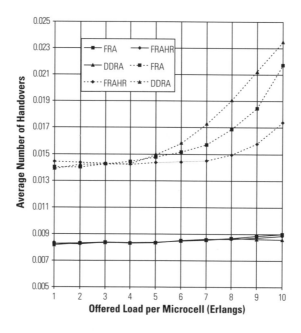

Figure 7.17 Average number of handovers an engaged call is expected to do during its lifetime. The dashed lines represent fast-moving users and solid lines represent slow-moving users. ©1999 IEEE. Reprinted, with permission, from [8].

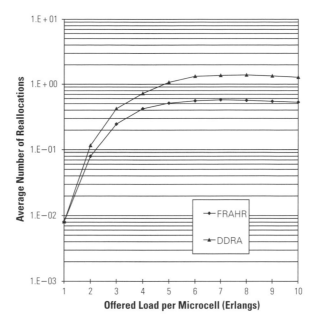

Figure 7.18 Average number of channel reallocations an engaged call might experience during its lifetime in the HCS with different resource assignments schemes. ©1999 IEEE. Reprinted, with permission, from [8].

7.7 Summary

This chapter has evaluated a new DRA strategy suitable for operation in a HCS. This strategy uses channel reallocation, hand-up, and hand-down to decrease the blocking probability in the system as well as the forced termination probability. The DDRA only requires that each BS has limited information about resource usage, thus making this strategy distributed. This last feature decreases the communication between BSs and therefore the signaling load in the system when compared with centralized DRA strategies. This strategy has been successfully evaluated by means of simulations using a Hybrid TD/ CDMA system proposed for future UMTS called FM1. From the simulation results, it can be said that:

1. DDRA performs much better than the FRA strategy and even better than the FRA strategy with HDP and channel reallocations.

2. The forced termination probability (and therefore the number of dropped calls) in the HCS is significantly smaller when using DDRA

rather than FRA. This is true for all types of service studied in this work and for all types of user.

3. Although service type 5 experiences a high blocking probability, different figures for this parameter could be obtained using more resources in the evaluation or a different number of microcells overlaid by each macrocell. This may change when using TDD (symmetric and asymmetric).

4. Basically, all types of service can be held in the HCS with the system proposed by ETSI (FM1). However, if the ratios of users requesting each type of service changes, then performance results could vary considerably.

Implementation of DDRA is worthwhile because there are enough advantages in doing so, such as an increase of the load the system can carry to meet a specific quality of service, and reduction of dropped calls, reduction of signaling load compared to other DCA strategies. Although we have tried to consider as many parameters as possible in the evaluation to make simulations similar to real case conditions, results may change in practical systems due to the limitations of the computer simulation model. One major omission is that only circuit-switched services have been taken into account, which is not a real case. For future work, packet-switched services must also be considered and their impact on the DDRA should be investigated to add new characteristics that can cope with these types of service.

References

[1]   ETSI STC SMG2#20, "FRAMES Multiple Access Scheme Proposal for the UMTS Radio Interface-SMG2 Workshop on UMTS Radio Interface Technologies," Sophia Antiopolis, Dec. 16–20, 1996.

[2]   Delli Priscoli, Francesco, et al., "Application of Dynamic Channel Allocation Strategies to the GSM Cellular Network," *IEEE Trans. Vehicular Tech.*, Vol. 15, No. 8, Oct. 1997, pp. 1558–1567.

[3]   Katzela, I., and M. Naghshineh, "Channel Assignment Schemes for Cellular Mobile Telecommunication Systems: A Comprehensive Survey," *IEEE Personal Commun. Mag.*, Vol. 3, No. 3, June 1996, pp. 10–31.

[4]   Scheibenbogen, Markus, et al., "Dynamical Channel Allocation in Hierarchical Cellular Systems," in *Proc. IEEE Vehicular Technol. Conf. VTC'96*, pp. 721-25, Atlanta, GA, USA, April 28–May 1, 1996.

[5] Lo, Kuen-Rong, et al., "A Combined Channel Assignment Strategy in a Hierarchical Cellular Systems," in *Proc. IEEE Int. Conf. on Universal and Personal Commun.*, *ICUPC'97*, pp. 651–655, San Diego, CA, USA, Oct. 1997.

[6] Klein, Anja, et al. "FRAMES Multiple Access Mode 1–Wideband TDMA With and Without Spreading," in *Proc. IEEE PIMRC'97*, pp. 37–41, Helsinki, Finland, Sept. 1997.

[7] Ortigoza G., Lauro, and A. H. Aghvami, "On the Optimum Spectrum Partitioning in a Microcell/Macrocell Cellular Layout with Overflow," in *Proc. IEEE GLOBECOM 97*, pp. 991–995,Phoenix, AZ, USA, Nov. 1997.

[8] Ortigoza-Guerrero, Lauro, and A. H. Aghvami, "A Distributed Dynamic Resource Allocation for a Hybrid TDMA/CDMA System," *IEEE Transactions on Vehicular Technology*, Vol. 47, No.4, pp. 1162-1178, Nov. 1998.

[9] Ovesjo, Frederik, et al., "Frames Multiple Access Mode 2- Wideband CDMA," in *Proc. IEEE, PIMRC'97*, pp. 42–46, Helsinki, Finland, Sept. 1997.

[10] ETSI SMG2, "Wideband TD/CDMA, Evaluation Report–Part 2," 1997.

[11] Frullone, M., G. Riva, P. Grazioso, and C. Carciofi, "Analysis of Optimum Resource Management Strategies in Layered Cellular Structures," in *Proc. Int. Conf. on Universal and Personal Commun., ICUPC'94*, pp. 371–375, San Diego, CA, USA. 1994.

[12] UMTS 30.03, V3.0.0 (1997-05), DTR/SMG-50402 (3PC0032S.pdf), ETSI, SMG-5, "Universal Mobile Telecommunication System (UMTS); Selection Procedures for the Choice of Radio Transmission Technologies of the UMTS," Sophia Antiopolis Cedex, France, 1997.

[13] Pizzarroso, M., and J. Jimenez, "Common Basis for Evaluation of a TDMA and CODIT System Concepts," R2020/TDE/CA/DS/L/SIG5-1/al,25/08/95.

[14] De Hedouville, Isabelle, and Hedayat Azad, "Performance Evaluation of Multi-Rate Hybrid TDMA/CDMA Systems," internal Report, Center for Telecommunication Research, King's College London, Sept. 1997.

[15] Hu, Lon-Rong, and Stephen Rappaport, "Personal Communication Systems Using Multiple Hierarchical Cellular Overlays," *IEEE J. Select. Areas Commun.*, Vol. 13, No. 2, Feb. 1995, pp. 406–415.

[16] Lin, Yi Bing, Anthony R. Noerpel, and Daniel J. Harasty, "The Sub-Rating Channel Assignment Strategy for PCS Hand-Offs," *IEEE Trans. Vehicular Tech.*, Vol. 45, No. 1, Feb. 1996, pp. 122–129.

8

Overall Summary and Directions for Future Work

The issue of management and allocation of resources in multilayered cellular structures has been investigated in this book. Several CASs were proposed and/or modified and applied to single-layered systems first (microcellular systems) and then to two-layered HCSs. The topic of spectrum partitioning has also been addressed in this book when trying to optimize the division of resources in a two-layered HCS.

Throughout this book, several teletraffic analyses were performed either to model or to validate a proposed CAS. They were also a tool to develop new methods to partition the spectrum.

In view of the complexity involved, a set of assumptions in system functions (such as handovers and coverage) was made to facilitate the derivation of the formulas that describe the system in question. As expected, the complexity of the analysis was alleviated considerably with the simplifications, but many important parameters with great impact on the implementation of real cellular systems were not considered (i.e., irregular areas of coverage, imperfect power control, and imperfect user classification based on their speed).

As in the case of analysis, computer simulations were performed based on idealized systems. Simplifications are necessary and helpful but also contribute to create biased results, in some cases optimistic, and in some others pessimistic; that would not match a real system. Examples of these simplifications are environments formed by regular cell patterns (both microcells and macrocells), perfect hierarchical cellular structure scenarios, and simplified mobility and traffic models.

8.1   Further Research Required

The idealization of the simulation platforms and the systems analytically treated left some points uncovered that need further research. Much can be done still to produce more "adequate" results that match real life systems. Some of the possible improvements are discussed in the following subsections.

8.1.1   Packet-Switched Services

Packet-switched services will certainly form a vast part of the traffic offered to UMTS. In the near future, the traffic load created by data communications will overtake the traffic load created by voice communications. Therefore, it is important to take them into account when designing a CAS. Models for internet browsing and e-mail, for example, should be incorporated into simulations and analytical studies whenever possible. Of particular interest is the creation of a DDCA strategy with queuing techniques to cope with these services and circuit-switched services in a more real and complex HCS than that developed in this book. Several approaches could be proposed for the discipline of the queue. Queuing data will certainly introduce delay in the communications but will increase the throughput of the systems.

8.1.2.   Keeping Up to Date with the Modifications to UTRA-TDD

Since the parameters of the hybrid system WB-TD/CDMA are changing constantly as the standardization process for UTRA goes on, it will be necessary to update the capacity and spectrum efficiency assessment performed for this system in this book, taking TDD rather than FDD into consideration. A new study showing the potentialities of WB-TD/CDMA-TDD in a HCS is required to show the advantages of this system when asymmetric traffic is considered in the UL/DL and when a DCA rather than FCA strategy is being used.

A possible calculation of the spectrum efficiency for the UTRA-TDD mode could be performed and compared with the UTRA-FDD, evaluating them in the same environment and under the same conditions. Such a comparison would be beneficial to determine which mode is optimal for every service.

8.1.3   Resource Allocation in WB-CDMA/TDD

Throughout this book TDM-based systems have been preferred over the WB-CDMA system because they allow easy application of DCA strategies.

However, it would be profitable to spend time investigating the possibility of using a DCA strategy in the proposed, but not yet fully standardized, WB-CDMA-TDD system. A DCA will certainly increase system capacity and spectrum efficiency considerably.

8.1.4 Study of HCS with Different MAS in Each Layer

It would be worth considering the feasibility of the implementation and evaluation of a HCS using different MASs in different layers. For instance, it could be useful where the hybrid system WB-TD/CDMA-TDD in the microcellular layer is overlaid by a macrocellular layer using WB-CDMA-FDD. Using an asymmetric TDD in the microcell will certainly improve the system performance when carrying asymmetric traffic, and the large capacity offered by the CDMA-based system will make the macrocell layer an excellent overflow group. It will require, however, an analytical study of the interlayer interference.

Another area of great interest would be to consider using WB-CDMA-FDD and WB-CDMA-TDD in different layers of the hierarchy. Using a particular MAS in a particular layer should be studied carefully and the interference it will cause to the overlaying or overlaid system should be investigated.

8.1.5 Expanding the Teletraffic Analysis Presented

A point of interest not tackled in the teletraffic analysis for HCS in this book is the aspect of considering hand-down and hand-up in the calculations of spectrum partitioning and blocking probabilities. This addition may have more than a trivial degree of complexity particularly when a mixed service environment is considered. Every service in the system will require a different number of channels. This will become more complex yet, if packet circuit services are included in the analysis, as they will have to be soon.

It has to be said, however, that such analysis is not a simple task because the mean time services and service times of data traffic do not follow exponential distributions. This enormously complicates the teletraffic analysis.

8.1.6 Power Control

Perfect power control was always assumed in the evaluation of the CAS presented in this book. Different algorithms for power control should be considered in future evaluations to assess more accurately the capacity of the systems when using a particular CAS and/or MAS.

8.1.7. More Complex and Real Simulation Environments

In a system level simulation a user is said to be unsatisfied when one of the following conditions happens:

- There are no available resources to attend its call (blocked from the system).
- The CIR falls below a fixed threshold for more than a period of time τ.
- The CIR falls below a fixed threshold for more than $X\%$ of the total call duration.

In the capacity evaluations performed throughout this book, only the first condition was used to determine the QoS of the user. A more complete approach that takes into account the rest of the factors is needed to produce more accurate capacity values.

Not only is a good definition of QoS required but also the inclusions of complete propagation models that take into account fast fading. Novel techniques such as smart antennas could also be utilized to enrich the completeness of the evaluation.

Of course, as with several ideas in this chapter, simulation is likely to be much more complex. This places a limit on both what performance factors can be included, and whether real systems could actually include all these factors.

The authors hope that these suggestions, and indeed, the whole book, encourage others to carry on one aspect or another of research into this interesting topic, which is so vital and urgent to the success of universal third generation systems. Any helpful comment on anything in the book would be enthusiastically welcomed by us.

A

Derivation of the State Equations for Overflow Traffic and Macrocell Layer

A.1 Overflow Traffic Distribution

Once the blocking probability has been calculated for every type of service in the microcell layer, we proceed to calculate the mean and the variance of the overflow traffic by introducing a fictitious, infinite server overflow group as in [1] and as in Section 5.2.1, with the number of channels M sufficiently large compared with the load submitted to a microcell.

This system could be described by a state ($j_1, j_2, \ldots, j_X, N_1, N_2, \ldots, N_X$) with equilibrium state probability $Q(j_1, j_2, \ldots, j_X, N_1, N_2, \ldots, N_X)$ in which j_1 to j_X were defined earlier and N_i is the number of calls of type i present in the overflow group. Note that the calls served in a microcell have mobility, but the calls served by the fictitious group are stationary. This system is described in Figure A.1. The state equations are formulated next.

For the case when $0 \le k_1 < S$ and $0 \le Y < M$, where

$$k_1 = m_1 j_1 + m_2 j_2 + \cdots + m_X j_X \tag{A.1}$$

and

$$Y_1 = N_1 j_1 + N_2 j_2 + \cdots + N_X j_X \tag{A.2}$$

Defining

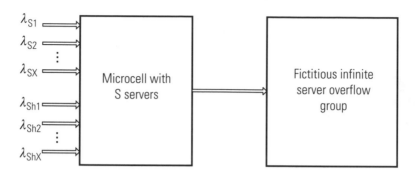

Figure A.1   Traffic that overflows from a microcell to a fictitious infinite server.

$$A_{1i} = \begin{cases} 1 & if \quad S - k_1 \geq m_i \\ 0 & \text{otherwise} \end{cases} \tag{A.3}$$

$$\left(\sum_{i=1}^{X} (\lambda_{Si} + \lambda_{Shi}) A_{1i} + \sum_{i=1}^{X} j_i \mu_i + \sum_{i=1}^{X} N_i \mu_i + \eta_{S1} \sum_{i=1}^{X} j_i \right) Q(j_1, j_2, \cdots, j_X, N_1, N_2, \cdots, N_X) \tag{A.4}$$

$$= (\lambda_{S1} + \lambda_{Sh1}) Q(j_1 - 1, j_2, \cdots, j_X, N_1, N_2, \cdots, N_X) +$$

$$+ (\lambda_{S2} + \lambda_{Sh2}) Q(j_1, j_2 - 1, \ldots, j_X, N_1, N_2, \cdots, N_X) + \cdots$$

$$+ (\lambda_{SX} + \lambda_{ShX}) Q(j_1, j_2, \cdots, j_X - 1, N_1, N_2, \ldots, N_X)$$

$$+ [\mu_1 (j_1 + 1) + \eta_{S1} (j_1 + 1)] Q(j_1 + 1, j_2, \ldots, j_X, N_1, N_2, \ldots, N_X) +$$

$$+ [\mu_2 (j_2 + 1) + \eta_{S1} (j_2 + 1)] Q(j_1, j_2 + 1, \ldots, j_X, N_1, N_2, \ldots, N_X) + \ldots$$

$$+ [\mu_X (j_x + 1) + \eta_{S1} (j_X + 1)] Q(j_1, j_2, \ldots, j_X + 1, N_1, N_2, \ldots, N_X) +$$

$$+ (N_1 + 1) \mu_1 Q(j_1, j_2, \ldots, j_X, N_1 + 1, N_2, \ldots, N_X) +$$

$$+ (N_2 + 1) \mu_2 Q(j_1, j_2, \ldots, j_X, N_1, N_2 + 1, \ldots, N_X) + \cdots$$

$$+ (N_X + 1) \mu_X Q(j_1, j_2, \ldots j_X, N_1, N_2, \ldots, N_X + 1)$$

The equilibrium equations for the boundary state $k_1 = S$ and $0 \leq Y \leq M$ are given next. For this purpose, let us define

$$A_2 = \begin{cases} 0 & if\ Y = M \\ 1 & otherwise \end{cases} \tag{A.5}$$

then

$$\left(\sum_{i=1}^{X} (\lambda_{Si} + \lambda_{Shi}) A_2 + \sum_{i=1}^{X} j_i \mu_i + \sum_{i=1}^{X} N_i \mu_i + \eta_{S1} \sum_{i=1}^{X} j_i \right) \tag{A.6}$$

$$\times Q(j_1, j_2, \dots, j_X, N_1, N_2, \dots, N_X)$$

$$= (\lambda_{S1} + \lambda_{Sh1}) Q(j_1, j_2, \dots, j_X, N_1 - 1, N_2, \dots, N_X) +$$

$$+ (\lambda_{S2} + \lambda_{Sh2}) Q(j_1, j_2, \dots, j_X, N_1, N_2 - 1, \dots, N_X) + \cdots$$

$$+ (\lambda_{SX} + \lambda_{ShX}) Q(j_1, j_2, \dots, j_X, N_1, N_2, \dots, N_X - 1)$$

$$+ (\lambda_{S1} + \lambda_{Sh1}) Q(j_1 - 1, j_2, \dots, j_X, N_1, N_2, \dots, N_X) +$$

$$+ (\lambda_{S2} + \lambda_{Sh2}) Q(j_1, j_2 - 1, \dots, j_X, N_1, N_2, \dots, N_X) + \cdots$$

$$+ (\lambda_{SX} + \lambda_{ShX}) Q(j_1, j_2, \dots, j_X - 1, N_1, N_2, \dots, N_X)$$

$$+ (N_1 + 1) \mu_1 Q(j_1, j_2, \dots, j_X, N_1 + 1, N_2, \dots, N_X) +$$

$$+ (N_2 + 1) \mu_2 Q(j_1, j_2, \dots, j_X, N_1, N_2 + 1, \dots, N_X) + \cdots$$

$$+ (N_X + 1) \mu_X Q(j_1, j_2, \dots, j_X, N_1, N_2, \dots, N_X + 1)$$

subject to the conditions $k_1 < S$ and

$$Q(j_1, \dots, j_i, \dots, j_X, N_1, N_2, \dots N_X) = 0 \ \forall \ j_i < 0 \tag{A.7}$$

$$Q(j_1, j_2, \dots, j_X, N_1, \dots, N_i, \dots, N_X) = 0 \ \forall \ N_i < 0 \ or \ N_i > M$$

The system is completely described by (A.4) and (A.6) and the following normalization equation

$$\sum_{j_1=0}^{S\,lm_1} \sum_{j_2=0}^{S\,lm_2} \cdots \sum_{j_X=0}^{S\,lm_X} \sum_{N_1=0}^{M\,lm_1} \sum_{N_2=0}^{M\,lm_2} \cdots \sum_{N_X=0}^{M\,lm_X} Q(j_1, j_2, \dots, j_X, N_1, N_2, \dots, N_X) = 1 \tag{A.8}$$

considering the summation of only the valid states for which $k_1 \leq S$.

The mean α_h and the variance v_h of the overflow distribution resulting from a subordinated microcell only can be computed from the state probabilities as [1]

$$\alpha_h = \sum_{j_1=0}^{Slm_1} \sum_{j_2=0}^{Slm_2} \cdots \sum_{j_X=0}^{Slm_X} \sum_{N_1=0}^{Mlm_1} \sum_{N_2=0}^{Mlm_2} \cdots \sum_{N_X=0}^{Mlm_X} N_i Q(j_1, j_2, \ldots, j_X, N_1, N_2, \ldots, N_X) \qquad (A.9)$$

$$(A.10)$$

$$v_h = \left\{ \sum_{j_1=0}^{Slm_1} \sum_{j_2=0}^{Slm_2} \cdots \sum_{j_X=0}^{Slm_X} \sum_{N_1=0}^{Mlm_1} \sum_{N_2=0}^{Mlm_2} \cdots \sum_{N_X=0}^{Mlm_X} N_i^2 Q(j_1, j_2, \ldots, j_X, N_1, N_2, \ldots, N_X) \right\} - \alpha_h^2$$

Then, remembering that each macrocell overlays exactly W microcells, the mean and variance of the composite overflow traffic could be found by

$$\alpha_T = \sum_{h=1}^{W} \alpha_h \qquad (A.11)$$

$$v_T = \sum_{h=1}^{W} v_h \qquad (A.12)$$

A.2 Overflow Traffic

The mean overflow traffic arrival rates from the microcells for new calls and hand-off calls of slow-moving users are given in terms of the IPP parameters discussed in Section 5.22

$$L_{Si} \approx \frac{(\gamma + \omega)}{\omega} (\lambda_{Si} Pb_i W) \qquad (A.13)$$

$$L_{Shi} \approx \frac{(\gamma + \omega)}{\omega} (\lambda_{Shi} Pb_i W) \qquad (A.14)$$

Once these parameters are determined, the states equations can be easily formulated by considering the case when the switch of the IPP is on and off.

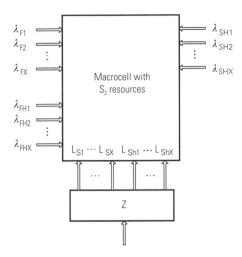

Figure A.2.   Macrocell under inspection.

A.3 Macrocell Layer

In each macrocell there are S_2 channels. An overlaying macrocell is described by the state $(j_1, j_2, \ldots, j_X, f_1, f_2, \ldots, f_X, Z)$ with equilibrium state probability $Q_2(j_1, j_2, \ldots, j_X, f_1, f_2, \ldots, f_X, Z)$, where $j_1, j_2, \ldots, j_X$ and $f_1, f_2, \ldots, f_X$ are the number of slow- and fast-moving users in the observed macrocell, respectively, for every type of service. Z is the state of the IPP random switch taking the value of 1 if the process is on or 0 if the process is off. The macrocell under inspection is shown in Figure A.2. The system equations are formulated in a similar way to the analysis shown for the microcell level. Let us define

$$A_{3i} = \begin{cases} 0 & if \ S_2 - k_2 = m_i \\ 1 & otherwise \end{cases}$$

where

$$k_2 = m_1 \left(j_1 + f_1 \right) + m_2 \left(j_2 + f_2 \right) + \cdots + m_X \left(j_X + f_X \right) \qquad (A.15)$$

The equilibrium state equations for $0 \le k_2 \le S_2$ are given by

$$\left(\sum_{i=1}^{X} \lambda_{Fi} A_{3i} + \sum_{i=1}^{X} \lambda_{FHi} A_{3i} + \sum_{i=1}^{X} \lambda_{SHi} A_{3i} + \sum_{i=1}^{X} f_i \mu_i + \right. \tag{A.16}$$

$$\left. + \sum_{i=1}^{X} j_i \mu_i + \eta_{F2} \sum_{i=1}^{X} f_i + \eta_{S2} \sum_{i=1}^{X} j_i + w \right) \cdot Q_2(j_1, j_2, \ldots, j_X, f_1, f_2, \ldots, f_X, 0)$$

$$= \gamma Q_2(j_1, j_2, \ldots, j_X, f_1, f_2, \ldots, f_X, 1)$$

$$+ (\lambda_{F1} + \lambda_{FH1}) Q_2(j_1, j_2, \ldots, j_X, f_1 - 1, f_2, \ldots, f_X, 0) +$$

$$+ (\lambda_{F2} + \lambda_{FH2}) Q_2(j_1, j_2, \ldots, j_X, f_1, f_2 - 1, \ldots, f_X, 0) + \cdots$$

$$+ (\lambda_{FX} + \lambda_{FHX}) Q_2(j_1, j_2, \ldots, j_X, f_1, f_2, \ldots, f_X - 1, 0) +$$

$$+ \lambda_{SH1} Q_2(j_1 - 1, j_2, \ldots, j_X, f_1, f_2, \ldots, f_X, 0) +$$

$$+ \lambda_{SH2} Q_2(j_1, j_2 - 1, \ldots, j_X, f_1, f_2, \ldots, f_X, 0) + \cdots$$

$$+ \lambda_{SHX} Q_2(j_1, j_2, \ldots, j_X - 1, f_1, f_2, \ldots, f_X, 0)$$

$$+ [(f_1 + 1)\mu_1 + (f_1 + 1)\eta_{F2}] Q_2(j_1, j_2, \ldots, j_X, f_1 + 1, f_2, \ldots, f_X, 0) +$$

$$+ [(f_2 + 1)\mu_2 + (f_2 + 1)\eta_{F2}] Q_2(j_1, j_2, \ldots, j_X, f_1, f_2 + 1, \ldots, f_X, 0) + \cdots$$

$$+ [(f_X + 1)\mu_X + (f_1 + 1)\eta_{F2}] Q_2(j_1, j_2, \ldots, j_X, f_1, f_2, \ldots, f_X + 1, 0) +$$

$$+ [(j_1 + 1)\mu_1 + (j_1 + 1)\eta_{S2}] Q_2(j_1 + 1, j_2, \ldots, j_X, f_1, f_2, \ldots, f_X, 0) +$$

$$+ [(j_2 + 1)\mu_2 + (j_2 + 1)\eta_{S2}] Q_2(j_1, j_2 + 1, \ldots, j_X, f_1, f_2, \ldots, f_X, 0) + \cdots$$

$$+ [(j_X + 1)\mu_X + (j_1 + 1)\eta_{S2}] Q_2(j_1, j_2, \ldots, j_X + 1, f_1, f_2, \ldots, f_X, 0)$$

with the constraints

$$Q_2(j_1, \ldots, j_i, \ldots, j_x, f_1, f_2, \ldots, f_x, 0) = 0 \qquad \forall \quad j_i < 0$$

and

$$Q_2(j_1, j_2, \ldots, j_x, f_1, \ldots, f_i, \ldots, f_x, 0) = 0 \qquad \forall \quad f_i < 0$$

Equation (A.16) is valid only for the cases for which $k_2 \leq S_2$

(A.17)

$$
\left(\sum_{i=1}^{X} \lambda_{Fi} A_{3i} + \sum_{i=1}^{X} \lambda_{FHi} A_{3i} + \sum_{i=1}^{X} \lambda_{SHi} A_{3i} + \sum_{i=1}^{X} L_{Si} A_{3i} + \sum_{i=1}^{X} L_{Shi} A_{3i} + \sum_{i=1}^{X} f_i \mu_i + \right.
$$

$$
\left. + \sum_{i=1}^{X} j_i \mu_i + \eta_{F2} \sum_{i=1}^{X} f_i + \eta_{S2} \sum_{i=1}^{X} j_i + \gamma \right) \cdot Q_2(j_1, j_2, \ldots, j_X, f_1, f_2, \ldots, f_X, 1)
$$

$$
= \omega Q_2(j_1, j_2, \ldots, j_X, f_1, f_2, \ldots, f_X, 0)
$$

$$
+ (\lambda_{F1} + \lambda_{FH1}) Q_2(j_1, j_2, \ldots, j_X, f_1 - 1, f_2, \ldots, f_X, 1) +
$$

$$
+ (\lambda_{F2} + \lambda_{FH2}) Q_2(j_1, j_2, \ldots, j_X, f_1, f_2 - 1, \ldots, f_X, 1) + \cdots
$$

$$
+ (\lambda_{FX} + \lambda_{FHX}) Q_2(j_1, j_2, \ldots, j_X, f_1, f_2, \ldots, f_X - 1, 1) +
$$

$$
+ (L_{S1} + L_{Sh1} + \lambda_{SH1}) Q_2(j_1 - 1, j_2, \ldots, j_X, f_1, f_2, \ldots, f_X, 1) +
$$

$$
+ (L_{S2} + L_{Sh2} + \lambda_{SH2}) Q_2(j_1, j_2 - 1, \ldots, j_X, f_1, f_2, \ldots, f_X, 1) + \cdots
$$

$$
+ (L_{SX} + L_{ShX} + \lambda_{SHX}) Q_2(j_1, j_2, \ldots, j_X - 1, f_1, f_2, \ldots, f_X, 1) +
$$

$$
+ \left[(f_1 + 1) \mu_1 + (f_1 + 1) \eta_{F2} \right] Q_2(j_1, j_2, \ldots, j_X, f_1 + 1, f_2, \ldots, f_X, 1) +
$$

$$
+ \left[(f_2 + 1) \mu_2 + (f_2 + 1) \eta_{F2} \right] Q_2(j_1, j_2, \ldots, j_X, f_1, f_2 + 1, \ldots, f_X, 1) + \cdots
$$

$$
+ \left[(f_X + 1) \mu_X + (f_X + 1) \eta_{F2} \right] Q_2(j_1, j_2, \ldots, j_X, f_1, f_2, \ldots, f_X + 1, 1) +
$$

$$
+ \left[(j_1 + 1) \mu_1 + (j_1 + 1) \eta_{S2} \right] Q_2(j_1 + 1, j_2, \ldots, j_X, f_1, f_2, \ldots, f_X, 1) +
$$

$$
+ \left[(j_2 + 1) \mu_2 + (j_2 + 1) \eta_{S2} \right] Q_2(j_1, j_2 + 1, \ldots, j_X, f_1, f_2, \ldots, f_X, 1) +
$$

$$
+ \left[(j_X + 1) \mu_X + (j_1 + 1) \eta_{S2} \right] Q_2(j_1, j_2, \ldots, j_X + 1, f_1, f_2, \ldots, f_X, 1) +
$$

with the constraints

$$
Q_2(j_1, \ldots, j_i, \ldots, j_x, f_1, f_2, \ldots f_x, 1) = 0 \qquad \forall \quad j_i < 0
$$

and

$$
Q_2(j_1, j_2, \ldots, j_x, f_1, \ldots, f_i, \ldots, f_x, 1) = 0 \qquad \forall \quad f_i < 0
$$

only for the valid cases for which $k_2 \leq S_2$.

The normalization equation in this case is

$$\sum_{j_1=0}^{S_2/m_1}\sum_{j_2=0}^{S_2/m_2}\cdots\sum_{j_X=0}^{S_2/m_X}\sum_{f_1=0}^{S_2/m_1}\sum_{f_2=0}^{S_2/m_2}\cdots\sum_{f_X=0}^{S_2/m_X}Q_2(j_1,\,j_2,\,\ldots,\,j_X,f_1,f_2,\ldots,f_X,\,1)=\frac{\omega}{\gamma+\varpi} \quad (A.18)$$

$$\sum_{j_1=0}^{S_2/m_1}\sum_{j_2=0}^{S_2/m_2}\cdots\sum_{j_X=0}^{S_2/m_X}\sum_{f_1=0}^{S_2/m_1}\sum_{f_2=0}^{S_2/m_2}\cdots\sum_{f_X=0}^{S_2/m_X}Q_2(j_1,\,j_2,\,\ldots,\,j_X,f_1,\,f_2,\ldots,\,f_X,0\,)=\frac{\gamma}{\gamma+\varpi} \quad (A.19)$$

The probability that a new call of any type is denied access to macrocell resources and hence blocked from the system is given by

$$P_{Mi} = \sum_{S_2-k_2<m_i}\Big[Q_2(j_1,\ldots,j_i,\ldots,j_X,f_1,\ldots,f_i,\ldots,f_X,0) \quad (A.20)$$
$$+ Q_2(j_1,\ldots,j_i,\ldots,j_X,f_1,\ldots,f_i,\ldots,f_X,1)\Big]$$

and the probability that the macrocell will not be able to serve a new call of any type when an overflowed slow-moving user call arrives is

$$P_{Bi} = \sum_{S_2-k_2<m_i}\,\underset{\forall i}{Q_2}(j_1,\ldots,\,j_i,\ldots,\,j_X,\,f_1,\ldots,\,f_i,\ldots,f_X,1) \quad (A.21)$$

Then the total blocking probability for a particular service i is given by

$$PBT_i = G_\mu pb_i\,P_{Bi} + G_M\,P_{Mi} \quad (A.22)$$

References

[1]   Hu, Lon-Rong, and Stephen Rappaport, "Personal Communication Systems Using Multiple Hierarchical Cellular Overlays," *IEEE J. Select. Areas Commun.*, Vol. 13, No. 2, Feb. 1995, pp. 406–415.

List of Acronyms

ACTS   Advanced communication technologies and services

BA   Basic algorithm

BAR   Basic algorithm with reassignment

BFA   Borrow first available

BCO   Borrowing with channel ordering

BDCL   Borrowing with directional channel locking

BS   Base station

BSC   Base station controller

CARB   Channel assignment with borrowing and reassignments

CAS   Channel allocation strategy

CCA   Combined channel assignment

CDMA   Code division multiple access scheme

CIR   Carrier to interference ratio

CPDCA Compact pattern-based dynamic channel allocation

CPMCB Compact pattern with maximized channel borrowing

CPMCB-NPS Compact pattern with maximized channel borrowing strategy with non-prioritized scheme for hand-offs

CPMCB-RCS1 Compact pattern with maximized channel borrowing strategy with reserved channel scheme with 1 reserved channel for hand-offs

CPMCB-RCS2 Compact pattern with maximized channel borrowing strategy with reserved channel scheme with 2 reserved channel for hand-offs

DCA Dynamic channel allocation

DDCA Distributed dynamic channel allocation

DDRA Distributed dynamic resource allocation

DH Directed hand-off

DL Downlink

DR Directed retry

DRA Dynamic resource allocation

DS Direct sequence

DTS Decision time slots

ETSI European Telecommunication Standard Institute

FA First available

FCA Fixed channel allocation

FCA-NPS Fixed channel allocation strategy with non-prioritized scheme for hand-offs

FCAR Fixed carrier allocation with reassignments

FCA-RCS1   Fixed channel allocation strategy with reserved channel scheme with 1 reserved channel for hand-offs

FCA-RCS2   Fixed channel allocation strategy with reserved channel scheme with 2 reserved channels for hand-offs

FDD   Frequency division duplex

FDMA   Frequency division multiple access

FIFO   First in first out

FMA   FRAMES multiple access

FMA1   FRAMES project Mode 1

FRA   Fixed resource allocation

FRAHR   Fixed resource allocation with hand-down and channel reallocations

FRAMES   Future radio wideband multiple access system

GOS   Grade of service

GSM   Global system for mobility

HCA   Hybrid channel allocation

HCS   Hierarchical cellular structure

HDCA-I   Hierarchical DCA at the carrier level

HDCA-II   Hierarchical DCA at the time slot level

HDP   Hand-down procedure

IPP   Interrupted Poisson process

LODA   Locally optimized dynamic assignment

LOS   Line of sight

MAHO   Mobile assisted handover

MAS   Multiple access scheme

MCHO   Mobile-controlled handover

MI   Minimum interference

MS   Mobile station

NCHO   Network controlled handover

NLOS   No line of sight

NPS   No priority scheme

ODCA   Ordered channel assignment scheme with rearrangement

ODMA   Opportunity-driven multiple access

OFDMA   Orthorgonal frequency division multiple access

PCS   Personal communications services

pdf   Probability density function

QAM   Quadrature amplitude modulation

QoS   Quality of service

QPSK   Quadrature phase shift keying

RAU   Remote antenna unit

RCS   Reserved channel scheme

RING   Selection with maximum usage on the reuse ring

RMI   Random minimum interference

RMIR   Random mimimum interference with reassignment

RNC   Remote network controller

SBR   Borrow from the richest

SDCA   Simple dynamic channel allocation

SHCB   Simple hybrid channel borrowing

SMG   Special mobile group

SMI   Sequential minimum interference

SNR   Signal-to-noise ratio

SRS   Subrating scheme

TDD   Time division duplex

TDMA   Time division multiple access scheme

UL   Uplink

UMTS   Universal mobile telecommunication systems

UTRA   UMTS terrestrial radio access

WB   Wideband

WB-TD/CDMA-TDD   Wideband time division multiple access scheme/ code division multiple access scheme with time division duplex

WB-CDMA-FDD   Wideband code division multiple access scheme with frequency division duplex

WCMTS   Wireless cellular mobile telecommunication system

List of Symbols

A_Ω Total offered load submitted to an area formed of a macrocell overlaying W microcells

a_μ Offered load to a single microcell

$a(k)$ Probability that a call makes a handover to one of its neighboring cells and is successfully accommodated

$a_{i,j}$ Offered loads produced by a service of types j

$a_{i,T}$ Total offered load submitted to a microcell i

a_M Offered load to a single macrocell

$b(k)$ Probability that a call currently served by cell k makes a handover attempt to a neighboring cell and is denied access to a channel in the target cell

c Probability that a user in cell k makes a handover

C/I Carrier to interference ratio

D The reuse distance

$D(k)$ Probability $D(k)$, that a call makes a handover after it has done zero, one,..., n previously successful handovers

$E(k)$ Probability that a mobile currently served by a cell in the system is forced to terminate after zero, one, ... , n successful handovers

$E\mu[v]$ The (conditional) mean velocities of mobiles in microcells

$E_M[v]$ The (conditional) mean velocities of mobiles in macrocells

$F(v)$ Velocity probability density function of mobiles

$f_{co}(t)$ Probability density function of the channel occupancy time distribution

L Path losses in the macrocellular layer

L_b Additional LOS path losses

L_{corner} Path losses caused by turning at a crossroad

L_{LOS} Line of sight path loss

L_{NLOS} No line of sight path loss

N Reuse pattern in a cellular system

pb Blocking probability

Pb_1 Probability that a new call from a *slow-moving user* is blocked from the microcell and thus overflowed to the overlaying macrocell

Pb_2 The probability that a new call (of any type) is denied access to a macrocell and thus blocked from the hierarchical structure

PB_2 The conditional probability that a macrocell will not be able to serve a new call when a overflowed slow-moving user call arrives

Pb_{ST} The total blocking probability for a new call origination from a slow moving user

P_{bT} Overall blocking probability

P_{ftt} Overall forced termination probability

ph Handover failure probability

Ph_1 Probability that a handover call from a *slow-moving user*, is overflowed to the overlaying macrocell

Ph_2 The probability that a handover call from a slow-moving user is denied access to a channel in a macrocell and hence forced to terminate

PH_2 The conditional probability that the macrocell will not be able to serve a handover call when a overflowed *slow moving user* handover call arrives

$P_i(j_i)$ Probability that in a system there are j users of the type of service i

P_j Probability of the state j in a group of servers

$p_{nc}(k)$ Probability that a call is not completed in cell k

R_b Breakpoint in the propagation model for microcellular systems

$R_k(s_k)$ Blocking rate in cell k as a function of the number of channels s_k

S Number of channels per cell

$S(k)$ Probability that a new call origination is accommodated by the system in cell k

vi The variance of the overflow distribution resulting from a subordinated cell only

Z Compact pattern with larger reduction of the overall blocking probability

α_i The mean of the overflow distribution resulting from a subordinated cell only

η_{NLOS} No line of sight coefficient

$\lambda_{hi}(k)$ Handover arrival rate from cell k

$\lambda_{ho}(k)$ Handover departure rate from cell k

λ_m Offered traffic to a macrocell

λ_T Total offered load to an *area* in microcellular system

$\lambda\mu$ Offered traffic to a microcell

τ_o Dwell time threshold

$\Omega_k(s_k)$ Blocking rate in cell k as a function of the number of channels s_k

About the Authors

Lauro Ortigoza-Guerrero received a Bachelor of Science degree in Electronic and Communications Engineering from ESIME-UPC in 1993 and the Masters of Science degree in Electrical Engineering from CINVESTAV in 1996, both from Instituto Politécnico Nacional in Mexico City, Mexico. He is currently working toward a Ph.D. at the Center for Telecommunications Research at King's College London, United Kingdom.

His area of interest is mobile cellular communication networks with a concentration on dynamic channel allocation strategies in mobile cellular systems with hierarchical cellular structures using hybrid multiple access schemes for UMTS, traffic analysis, and performance evaluation.

Professor A. Hamid Aghvami obtained his Masters of Science and Ph.D. degrees from King's College, The University of London, in 1978 and 1981, respectively. In April 1981 he joined the department of Electronic and Electrical Engineering at King's as a postdoctoral research associate. He worked on digital communications and microwave techniques projects sponsored by EPSRC. He joined the academic staff at King's in 1984. In 1989 he was promoted to Reader and to Professor in Telecommunications Engineering in 1992. He is presently the director of the Center for Telecommunications Research at King's College. Professor Aghvami performs consulting work on Digital Radio Communication Systems for both British and International companies. He has published over 200 technical papers and lectures on Digital Radio Communications, including GSM 900/DCS 1800 worldwide. He leads an active research team working on numerous mobile and personal communication system projects for third generation systems; these projects are supported both by the government and industry. He is a distinguished lecturer of the IEEE Communications Society. He has been member, Chairman, and Vice-Chairman of the technical program and organizing committees of a large number of international conferences. He is also the founder of the International Conference on Personal Indoor and Mobile Radio Communications. He is a Fellow member of the IEE and a senior member of the IEEE.

Index

Recent Titles in the Artech House
Mobile Communications Series

John Walker, Series Editor

For further information on these and other Artech House titles, including previously considered out-of-print books now available through our In-Print-Forever® (IPF®) program, contact:

Artech House
685 Canton Street
Norwood, MA 02062
Phone: 781-769-9750
Fax: 781-769-6334
e-mail: artech@artechhouse.com

Artech House
46 Gillingham Street
London SW1V 1AH UK
Phone: +44 (0)20 7596-8750
Fax: +44 (0)20 7630-0166
e-mail: artech-uk@artechhouse.com

Find us on the World Wide Web at:
www.artechhouse.com